T0180470

Texts and Monographs in
Symbolic Computation

A Series of the
Research Institute for Symbolic Computation,
Johannes-Kepler-University, Linz, Austria

Edited by
B. Buchberger and G. E. Collins

N. Kajler (ed.)

Computer–Human Interaction in Symbolic Computation

With a Foreword
by D. S. Scott

SpringerWienNewYork

Dr. Norbert Kajler
Centre de Calcul
Ecole des Mines de Paris, Paris, France

Data conversion by Thomson Press (India) Ltd., New Delhi, India
Printed by Novographic, Ing. Wolfgang Schmid, A-1230 Wien
Graphic design: Ecke Bonk
Printed on acid-free and chlorine-free bleached paper

With 68 Figures

Library of Congress Cataloging-in-Publication Data

Computer-human interaction in symbolic computation / N. Kajler, ed.:
with a foreword by Dana S. Scott.
p. cm. — (Texts and monographs in symbolic computation, ISSN 0943-853X)
Includes bibliographical references and index.
ISBN 3-211-82843-5 (alk. paper).
1. Human-computer interaction. I. Kajler, N. (Norbert) II. Series.
QA76.9.H85C657 1998 97-42077
005'.01'9—DC21 CIP

ISSN 0943-853X
ISBN 3-211-82843-5 Springer-Verlag Wien New York

Foreword

The well attended March 1994 HISC workshop in Amsterdam was a very lively conference which stimulated much discussion and human–human interaction. As the editor of this volume points out, the Amsterdam meeting was just part of a year-long project that brought many people together from many parts of the world. The value of the effort was not only in generating new ideas, but in making people aware of work that has gone on on many fronts in using computers to make mathematics more understandable. The author was very glad he attended the workshop.*

In thinking back over the conference and in reading the papers in this collection, the author feels there are perhaps four major conclusions to be drawn from the current state of work:

1. graphics is very important, but such features should be made as easy to use as possible;
2. symbolic mathematical computation is very powerful, but the user must be able to see "intermediate steps";
3. system design has made much progress, but for semester-long coursework and book-length productions we need more tools to help composition and navigation;
4. monolithic systems are perhaps not the best direction for the future, as different users have different needs and may have to link together many kinds of tools.

The editor of this volume and the authors of the papers presented here have also reached and documented similar conclusions.

As regards graphics, if, as the old saying goes,"a picture is worth a thousand words," then a good picture is worth several thousand, and animated pictures have an incalculable value. There are many video films now available that show vividly how good graphics and animation can substantially aid mathematical intuition. Also, the recent advances in workstation design and software products have brought to the individual user the power for generating computer graphics that used to require hours of supercomputer use. But, people have to be taught to use such tools, and they have to be warned about pitfalls. The approximations done in graphics programs can create false pictures and strange, irrelevant graphical artifacts. Authors have to remember to give us collections of bad examples as well as collections of good examples. Indeed, understanding why a "numerical breakdown" creates a bad example can be a very informative experience. Linking graphics to symbolic manipulation is also an encouraging development, since the user or the student needs to think how the picture is defined. Being able to quickly

*A revised version of the author's contribution to the workshop on teaching with Mathematica appeared in *Lecture Notes in Computer Science*, vol. 1138, edited by J. Calmet, J.A. Campbell, and J. Pfalzgraf (Springer, Berlin Heidelberg New York Tokyo).

produce or edit examples where the equations (or other formulae) can be controlled by the user is a great aid to learning and experimentation.

As regards symbol manipulation, the author or programmer of interesting examples must remember that even though he or she knows how the solution is obtained, a later reader or student may have a very imperfect understanding or a wrong mental model of processes invoked. Even a simple algebraic transformation may entail hundreds of operations in the computer. The objective of learning to use the computer is not to rerun canned examples, but to be able to solve new problems. Thus, examples have to be modifiable and explainable. System builders also have to be aware that usable "trace" programs are needed to debug examples. It is also key to be able to program individually new symbol manipulations that are needed for new algorithms or new applications. The "perfect" design of a symbol manipulation language has perhaps not yet been found.

As regards system design in general, the designer has to remember that materials using the system have to grow and evolve. Moreover, new users have to be able to find information easily. When there is only one demonstration to give or only a few computer files involved, then it is fairly easy to make the setup understandable. But a long-term multi-user or multi-author project needs reliable file management and – especially – indexing and searching. More and more university courses in the U.S. are now using the World-Wide Web, which makes often a simple "point-and-click" interface to an archive of files. But more thought has to be given to interfacing WWW pages to software and to providing the security that is needed for students doing coursework.

As regards modularity, we have to remember that over several years, application programs are going to change and materials from many different sources will have to be combined. Upward compatibility for new versions of systems, and cross-platform or inter-program compatibility of source files has to be taken into account. Also, a general-purpose system can never satisfy all special-purpose users. Thus, different components have to be able to interact (and the expanding use of the WWW makes that even more urgent). Another problem is also acute for symbolic mathematical computation: automatic-proof systems have to be linked to computer-algebra systems. In the present state of development this is not at all easy to do, and it is hardly at a phase where combined systems could be used for education on a large scale. Someone has to come up with some good proposals on how "plug-and-play" systems can be built and how they can be fairly marketed. For some areas of business it may be acceptable to have "closed-world" system installations, but for education and research it is not reasonable to build many walls. Compatibility and expandability are not easily achievable, however.

The issues briefly touched on above are often lumped together under the heading "hypertext." The present writer does not feel that we understand hypertext and hypermedia sufficiently yet to say it has really provided solutions. There are many good ideas around, but more integration is needed. The problems of using symbolic mathematical computation and turning much of our standard introductory-school and college mathematics curriculum into computer-based resources is a great testing ground for sparking new developments. It is good to see that there are a number of interesting projects doing just this.

Dana S. Scott
Carnegie Mellon University,
Pittsburgh, Pa.

Contents

List of contributors

Theodore Alper, EPGY, Ventura Hall, Stanford University, Stanford, CA 94305-4115, U.S.A.

Kostadin Antchev, Hypermedia Laboratory, Department of Mathematics, Tampere University of Technology P.O. Box 692, FIN-33101 Tampere, Finland.

Ron Avitzur, RIACA, Eindhoven, The Netherlands. Permanent address: P.O. Box 6692, Stanford, CA 94309, U.S.A.

Olivier Arsac, Projet SAFIR, INRIA Sophia Antipolis, 2004 route des lucioles, B.P. 93, F-06902 Sophia-Antipolis Cedex, France.

Olaf Bachmann, RIACA, Eindhoven, The Netherlands. Permanent address: Department of Mathematics and Computer Science, Kent State University, Kent, OH 44242, U.S.A.

Michael Beeson, Department of Mathematics and Computer Science, San Jose State University, San Jose, CA 95192, U.S.A.

Arjeh M. Cohen, RIACA, Faculteit Wiskunde en Informatica, Technische Universiteit Eindhoven, Postbus 513, 5600 MB Eindhoven, The Netherlands.

Stéphane Dalmas, Projet SAFIR, INRIA Sophia Antipolis, 2004 route des lucioles, B.P. 93, F-06902 Sophia-Antipolis Cedex, France.

Maylis Delest, LaBRI, Unité associée au C.N.R.S. no. 1304, Université Bordeaux I, F-33405 Talence Cedex, France.

Jean-Marc Fédou, LaBRI, Unité associée au C.N.R.S. no. 1304, Université Bordeaux I, F-33405 Talence Cedex, France.

Marc Gaëtano, Projet SAFIR, INRIA Sophia Antipolis, 2004 route des lucioles, B.P. 93, F-06902 Sophia-Antipolis Cedex, France.

Norbert Kajler, RIACA, Eindhoven, The Netherlands. Permanent address: Ecole des Mines de Paris, 60 boulevard St-Michel, F-75006 Paris, France.

Lambert Meertens, Centrum voor Wiskunde en Informatica, P.O. Box 94079, 1090 GB Amsterdam, The Netherlands.

Guy Melançon, LaBRI, Unité associée au C.N.R.S. no. 1304, Université Bordeaux I, F-33405 Talence Cedex, France.

Jari Multisilta, Hypermedia Laboratory, Department of Mathematics, Tampere University of Technology, P.O. Box 692, FIN-33101 Tampere, Finland.

Jean Paoli, Grif S.A., 2 boulevard Vauban, B.P. 266, F-78053 St. Quentin en Yvelines, France.

Seppo Pohjolainen, Hypermedia Laboratory, Department of Mathematics, Tampere University of Technology, P.O. Box 692, FIN-33101 Tampere, Finland.

Vincent Quint, INRIA Rhone-Alpes, 655 avenue de l'Europe, F-38330 Montbonnot St. Martin, France.

Ray Ravaglia, EPGY, Ventura Hall, Stanford University, Stanford, CA 94305-4115, U.S.A.

Nadine Rouillon, LaBRI, Unité associée au C.N.R.S. no. 1304, Université Bordeaux I, F-33405 Talence Cedex, France.

Marianna Rozenfeld, EPGY, Ventura Hall, Stanford University, Stanford, CA 94305-4115, U.S.A.

Patrick Suppes, EPGY, Ventura Hall, Stanford University, Stanford, CA 94305-4115, U.S.A.

Irène Vatton, INRIA Rhone-Alpes, 655 avenue de l'Europe, F-38330 Montbonnot St. Martin, France.

Introduction

Norbert Kajler

The goal of the project Human Interaction in Symbolic Computing (HISC) which took place in 1994–1995 at the Research Institute for Applications of Computer Algebra (RIACA) in Amsterdam was to investigate a variety of techniques and paradigms which could lead to better user interfaces to symbolic-computation systems (current and future).

There are indeed many problems which current user interfaces either do not handle well or do not address at all. These include the efficient presentation and editing of possibly large mathematical expressions; use of alternative input technologies, such as voice or pen; direct manipulation of subexpressions at both the syntactic and semantic levels; interactive manipulation of curves and surfaces; representation and interactive manipulation of exotic mathematical objects; intelligent session management providing support to users solving nontrivial problems; interaction with theorem provers/checkers, etc.

At this point, the research undertaken during the HISC project concentrated on a limited number of problems, namely: curve and surface plotting, visualization of mathematical objects, data exchange between computer algebra kernels and user interfaces, and interactive books.

As with the two other projects initiated at RIACA in 1994, Optics and Dynamical Systems, several researchers visited RIACA for varying lengths of time. Among them were: Simon Atkins (University of Bath, U.K.), Ron Avitzur (U.S.A.), Olaf Bachmann (Kent State University, U.S.A.), Thomas Banchoff (Brown University, U.S.A.), Simon Gray (Kent State University, U.S.A.), Michael Mac Gettrick (Waterford Regional Technical College, Ireland), Jari Multisilta (Tampere University of Technology, Finland), Nadine Rouillon (Université de Bordeaux I, France), Neil Soiffer (Wolfram Research, U.S.A.), and Eugene Zima (Moscow State University, Russia).

In addition to this volume, the HISC project resulted in several publications with the most visible being a survey of user interfaces for computer algebra systems (Kajler and Soiffer 1998) and the following articles: Gray et al. (1994); Fournier et al. (1995); Avitzur et al. (1995), with additional reports and preprints available in the RIACA technical report series.

A more complete report of the HISC project can be found in Kajler (1995).

Beyond the work done during the HISC project, Human Interaction in Symbolic Computing is a wide and interdisciplinary research area whose main connections include symbolic and algebraic computing, computer–human interaction, scientific visualization, document processing, and computer-aided education.

The problem of providing convenient user interfaces for symbolic computation tools was first stated quite some time ago. As early as 1963, Minsky presented in the MATH-

SCOPE proposal some general directions for the manipulations of mathematical expressions on a screen. Since then, many meetings and publications contributed to a more precise understanding of how user interfaces for symbolic-computation tools could be improved (Wells and Morris 1972, Foster 1984, Arnon 1987, Katz 1987, Soiffer 1991, Kajler 1993). Effective implementations took more time to appear, as can be seen in the history section of Kajler and Soiffer (1998).

Nowadays, a number of long-awaited user interface features are becoming more available in the area of symbolic computation software. These include two-dimensional display of mathematical expressions with the ability to point at and edit subexpressions; the possibility to mix text, formulas, and graphics, within a single paper-like graphical window; straightforward access to various forms of on-line documentation including tutorial, reference manual, and examples.

Still, some other desirable capabilities are only offered by a few systems or prototypes. For example, direct manipulation of mathematical expressions – as introduced with Milo (Avitzur 1988) and refined in Theorist (Bonadio 1989) – is not yet very popular among mainstream systems. The same is true for extensibility of the user interface, including the addition of new menus, new graphical notations, new pieces of on-line documentation, etc. Most recent advances along these lines include CAS/PI (Kajler 1992) and the new Mathematica typesetter (Soiffer 1995). Also, few user interfaces support concurrent use of different symbolic-computation packages, dealing with conversion problems, configuration management, and communication protocols. Pioneering work on these subjects was done by James Purtilo (1985) with Polylith and Dennis Arnon et al. (1988) with CaminoReal. More recent results include SUI (Doleh and Wang 1990) and CAS/PI (Kajler 1992).

Clearly, all these features are far from trivial to implement and, furthermore, cause additional difficulties in terms of the coherence and ergonomy of the resulting interface.

Concerning coherence and ergonomy, more efforts could be done in order to improve upon the existing mechanisms and paradigms used to display and interact with especially large expressions on the screen. The same is true for session layout which does not necessarily have to be a sequence of input–output entries mingled in a single window. The same is also true for computer-aided education software in which interactivity and ease of use should certainly be maximized. Still, only costly and time-consuming usability testing can validate improved user interaction models that may replace existing ones.

Other open research areas include the use of symbolic-computation software via a WWW browser (with efficient communication of possibly large datasets and high-quality display of the equations) and the effective use of alternative input technologies such as pen or voice which could complement or supplement the traditional mouse and keyboard devices (at least in a number of specific situations – visually impaired and physically challenged users for instance).

Consequently, research in the field of HISC should start moving away from the design and implementation of complete user interface prototypes such as Mathscribe (Smith and Soiffer 1986), GI/S (Young and Wang 1987), CaminoReal (Arnon et al. 1988), or CAS/PI (Kajler 1992), toward more focused experiments. Our vision of truly productive user interface work in the area of HISC is one of research focused on specific problems, with testing and validation of carefully tuned interaction techniques, and the implementation of separate *re-usable* software components to be loaded into some larger modular construction.

Most of the contributors to this volume participated in the HISC project and/or the HISC workshop organized on March 10–11, 1994 in Amsterdam. Two additional authors, Beeson and Ravaglia, were solicited by the editor in order to offer insights from the point of view of computer-aided education.

The organization of this book follows a three-part pattern, with each part composed of three papers.

The first part presents the "interactive book" perspective; that is, how to provide users with an easy-to-use, yet powerful, means of interaction with symbolic-computation packages that is based on the book metaphor. The first paper in this part elaborates on the need for requirements through the presentation of the ACELA project. The aim of the ACELA project is to develop a framework for authoring and browsing interactive mathematical books, and also to complete a series of interactive books starting with a volume on Lie algebras. The second paper is about the software architecture issues behind the development of interactive books. The possibilities offered by the Grif technology are presented in addition to the advantages of a high level of abstraction in the way mathematical information is stored and exchanged. The final paper in the first part highlights the new possibilities for the direct manipulation of equations and graphs allowed by powerful, modern desktop computers. Mouse-based direct manipulation of equations and graphs appears to be a wonderful exploratory tool, mostly desirable from a pedagogical point of view.

The second part presents three experiments in the field of computer-aided instruction which have in common an extensive use of symbolic-computation techniques. The first two papers in this part describe different ways to achieve a similar goal: how to use symbolic-computation techniques in the most pedagogically effective way. In the first paper, the authors describe how they developed a sophisticated framework around an existing CAS package (Maple V). Using an EPGY computer course, students usually watch a pre-recorded multimedia lesson and, later on, test their understanding by solving some exercises under the supervision of the computer. In the second paper, the author advocates that the best way to provide a pedagogically satisfying mathematical tutor was to design a brand new system based on some specific requirements. The presentation of the Mathpert assistant follows. The third paper in this part presents HMLE, a hypermedia-based learning environment, and focuses on the possible benefits of the combined uses of numeric and symbolic packages and multimedia techniques in the area of computer-aided instruction.

The third part groups three papers related to visualization. The first paper presents the new possibilities offered by the "chains of recurrences" technique to expedite the visualization and animation of explicitly defined curves and surfaces. The technique is based on the automatic generation of chains of recurrence relations, given some closed-form functions to be evaluated repeatedly over a regular grid. The second paper deals with the visualization of the behavior of algorithms and presents the Agat software package which was designed to facilitate the task of setting up new algorithm animations. The last paper presents CallCo, an on-going project at the edge of symbolic computation, distributed computing, and mathematical visualization whose aim is to investigate how best to display and interact with combinatorial objects.

We hope that this book will be of value to those people involved in the design and implementation of scientific software, especially symbolic-computation packages. They should be able to draw upon the research material presented in this book to create more powerful and user-friendly applications for the benefit of all "end-users". Equally, we hope that the book will encourage researchers to push forward their investigations in the area of HISC and to continue to explore the many open and challenging aspects it has to offer.

Acknowledgments

The preparation of this book as well as part of the work reported herein was supported by the Research Institute for Applications of Computer Algebra (RIACA). Additional support for the preparation of this volume was provided by the University of Nice-Sophia Antipolis and the Ecole des Mines de Paris.

I would like to thank the series editors and Springer-Verlag Wien for offering to publish this collection of articles as part of the series *Texts and Monographs in Symbolic Computation*. I also would like to thank all the authors who submitted their manuscripts for publication in this volume, all the participants in the HISC project and workshop, and all the referees who assisted in the reviewing of the papers, namely: Isabelle Attali, Ron Avitzur, Thomas F. Banchoff, Frédérique Barkats, Monique Baron, Bernard Beauzamy, Yves Bertot, Marc Brown, Ed Clarke, Arjeh Cohen, Stéphane Dalmas, James Davenport, Sam Dooley, Paul Drijvers, Frédéric Eyssette, Richard Fateman, Marc Gaëtano, Arnaud Gourdol, Andrew Hanson, Lynda Hardman, André Heck, Frans Heeman, Pieter Hemker, Karsten Homann, Hoon Hong, Paul Jackson, Manfred Kerber, Paul Klint, Franz Kriftner, Silvio Levy, Michael MacGettrick, Bernard Mourrain, Jean-François Nicaud, Peter Paule, Steven Pemberton, Frank Pfening, Didier Pinchon, Gerald Porter, Loïc Pottier, Sarah Rees, Cecile Roisin, Carol Scheftic, Fred Simons, Steven Skiena, Gerard Sleijpen, Neil Soiffer, John Stasko, Robert Sutor, Laurent Théry, Annick Vallibouze, Hans van Hulzen, Marc van Leeuwen, Robert van Liere, Bjorn van Sydow, C. R. Traas, Jim Welsh, Eugene Zima.

References

Arnon, D. (1987): Report of the Workshop on Environments for Computational Mathematics, held July 30, 1987, during the ACM SIGGRAPH Conference. ACM SIGSAM Bull. 21/4: 42–48.

Arnon, D., Beach, R., McIsaac, K., Waldspurger, C. (1988): CaminoReal: an interactive mathematical notebook. In: van Vliet, J. C. (ed.): Proceedings of EP '88 International Conference on Electronic Publishing, Document Manipulation, and Typography. Cambridge University Press, Cambridge, pp. 1–18.

Avitzur, R. (1988): Milo (a Macintosh computer program). Paracomp Inc. San Francisco.

Avitzur, R., Bachmann, O., Kajler, N. (1995): From honest to intelligent plotting. In: Levelt, A. H. M. (ed.): Proceedings of the International Symposium on Symbolic and Algebraic Computation (ISSAC '95), Montreal, Canada. Association for Computing Machinery, New York, pp. 32–41.

Bonadio, A. (1989): Theorist (a computer program). Prescience Corp., San Francisco.

Doleh, Y., Wang, P. S. (1990): SUI: a system independent user interface for an integrated scientific computing environment. In: Proceedings of the International Symposium on Symbolic and Algebraic Computation (ISSAC '90), Tokyo, Japan. Addison-Wesley, Reading, MA, pp. 88–94.

Foster, G. (1984): User interface considerations for algebraic manipulation systems. Techn. Rep. UCB/CSD-84-192, University of California, Berkeley, CA

Fournier, R., Kajler, N., Mourrain, B. (1995): Visualization of mathematical surfaces: the IZIC server approach. J. Symb. Comput. 19: 159–173.

Gray, S., Kajler, N., Wang, P. S. (1994): MP: a protocol for efficient exchange of mathematical expressions. In: Giesbrecht, M. (ed.): Proceedings of the International Symposium on Symbolic and Algebraic Computation (ISSAC '94), Oxford, U.K. Association for Computing Machinery, New York, pp. 330–335.

Kajler, N. (1992, July): CAS/PI: a portable and extensible interface for computer algebra systems. In: Wang, P. S. (ed.): Proceedings of the International Symposium on Symbolic and Algebraic Computation (ISSAC '92), Berkeley, U.S.A. Association for Computing Machinery, New York, pp. 376–386.

Kajler, N. (1993): Environnement graphique distribué pour le calcul formel. Thèse, Université de Nice-Sophia Antipolis, Ecole Doctorale SPI, Sophia Antipolis, France.

Kajler, N. (1995): The HISC project at RIACA. CAN Newslett. 14: 9–14.

Kajler, N., Soiffer, N. (1998): A survey of user interfaces for computer algebra systems. J. Symb. Comput. (to appear).

Katz, A. (1987): Issues in defining an equations representation standard. ACM SIGSAM Bull. May: 19–24.

Minsky, M. L. (1963): MATHSCOPE: part I – a proposal for a mathematical manipulation-display system. Techn. Rep. MAC-M-118, Artificial Intelligence Project, Project MAC, MIT, Cambridge, MA.

Purtilo, J. (1985): Polylith: an environment to support management of tool interfaces. In: Proceedings of the ACM SIGPLAN Symposium on Language Issues in Programming Environments. Association for Computing Machinery, New York, pp. 12–18.

Smith, C., Soiffer, N. (1986): MathScribe: a user interface for computer algebra systems. In: Char, B. (ed.): Proceedings of the 1986 Symposium on Symbolic and Algebraic Computation (SYMSAC '86), Waterloo, Canada. Association for Computing Machinery, New York, pp. 7–12.

Soiffer, N. (1995): Mathematical typesetting in Mathematica. In: Levelt, A. H. M. (ed.): Proceedings of the International Symposium on Symbolic and Algebraic Computation (ISSAC '95), Montreal, Canada. Association for Computing Machinery, New York, pp. 140–149.

Soiffer, N. M. (1991): The design of a user interface for computer algebra systems. Ph.D. thesis, University of California, Berkeley, CA.

Wells, M. B., Morris, J. B. (eds.) (1972): Proceedings of a Symposium on Two-Dimensional Man-Machine Communication. ACM SIGPLAN Notices 7.

Young, D. A., Wang, P. S. (1987): GI/S: a graphical user interface for symbolic computation systems. J. Symb. Comput. 4: 365–380.

The ACELA project: aims and plans

Arjeh M. Cohen and Lambert Meertens

1 Introduction

The most visible aim of the ACELA (architecture of a computer environment for Lie algebras) project is the production of a state-of-the-art interactive book on Lie algebras; state-of-the-art mathematically as well as in its interactive potential. While we have chosen this as a worthwhile and challenging goal by itself, this target also serves as a concrete milestone for our longer-term aims, offering a realistic and far from trivial testing ground for our ideas.

At the moment of writing, the project is still in an early stage. As it is primarily a research project – although having prototypes as deliverables – our ideas will undoubtedly continue to evolve as the work proceeds and influence the final product. So this paper should be read as a snapshot of our present views. Rather than giving an overview of all envisioned functionality, we focus on some of the aspects that we find most interesting. As we go along, we point out our various aims and plans.

The ACELA project is supported by the Netherlands Technology Foundation (STW), and, through the linked MathViews and WINST projects, by the Netherlands Foundation for Computer Science Research (SION) and the Netherlands Foundation for Mathematics (SMC). At present, some eight more people (Hugo Elbers, James Faux, Willem de Graaf, Nico van den Hijligenberg, André van Leeuwen, Marc van Leeuwen, Steven Pemberton, and Marcel Roelofs) work or did work directly on the project, with several others (Henk Barendregt, Michiel Hazewinkel, Norbert Kajler, and Jan-Willem Klop) involved more peripherally. It should be clear that the ideas presented here are not only those of the authors of this paper, but were shaped and refined by a collective effort.

2 Interactive books

Recent developments in information technology – hypertext techniques, multimedia facilities, CD-ROMs, networking – as well as the increasing affordability of sufficiently powerful computers with high-resolution graphical displays have greatly improved the possibilities of using computers for making knowledge accessible. Nevertheless, the prevalent way of learning about some advanced subject is still through the reading of books, whether or not aided by a human teacher who serves as an expert ready to explain those bits that you did not grasp from reading the book.

In general, that combination of text and expert is important. Traditional books are infinitely patient: you can consult them as often as you want. At the same time, they are oblivious to any need of a reader beyond the static text offered: you cannot pose questions to a book and expect an adequate response. Traditional teachers are not – or shouldn't be – oblivious to their students' needs, but infinitely patient they are definitely not. An interactive book can combine (some of) the best of both worlds, and at the same time encourage and support *active* participation of the reader – as opposed to the often too passive modes of reading a book or listening to a teacher.

The observation that human teachers are of limited patience has been one of the motivations for the earlier work on computer-aided instruction. We wish to point out a fundamental difference between "classical" CAI and our view of how the reader will interact with the book.

In classical CAI, the assumption is that there is a given syllabus, a well-delineated body of knowledge, that students are supposed to master, whether motivated or not. Accordingly, there is generally a fixed route through the material, with strong emphasis on testing if the student has indeed mastered a morsel of knowledge, typically with a dialogue controlled by the automated instructor.

In contrast, our assumption is that the reader, in consulting the book, *is* motivated to understand, or at least find out, something – and it is up to the reader to define *what* precisely that something is. It might be a mere trifling trinket in the wealth of wisdom stowed in the book. Thus, the *book* metaphor is essentially more apt than the *instructor* metaphor. In a paper book, the reader can just browse, instead of perusing it from cover to cover, consult it in any haphazard order, skip any sections, examples, exercises, et cetera, and come back to any point at any time. To call an interactive system an interactive *book*, not only should there be a text, prepared by an author for readers (this need not be text in the strictest sense of the word; the notion as we use it here also includes pictures of various kinds and in general organised information in any consultable form), but we also hold the tenet that *all interaction is initiated, and in general controlled, by the reader*.

Interactive books are not superior to more traditional books in all aspects. For the time being, they (or, rather, the equipment needed for consulting them) will be less easy to carry around. (On the other hand, for electronic books it hardly makes a difference whether we have a small text or a comprehensive encyclopedia.) The resolution of affordable screens is an order of magnitude less than that of high-quality printing, which makes it tiring to read the fine print in mathematical formulae like "n^{p_i}". Also, the initial investment in acquiring the necessary equipment is sizable, and the initial production cost is appreciably higher (although, on the other hand, the unit reproduction cost of the book texts themselves is basically the cost of the medium, and if distributed through networking virtually free).

A book's interactive potential has to be substantive before it offsets these disadvantages. We now discuss several possibilities that are offered by an interactive setup. Although we group them for convenience, the boundaries between these groups are not sharp. As a non-mathematical running example we take an interactive cookbook. Most of the sophisticated interactive "features" mentioned below for this cookbook can be found in one or more commercially available interactive-cookbook programs.

2.1 Navigation support

Navigation comprises "standard" hypertext facilities, which can be used both for unfolding/folding portions of a hierarchically structured text (as in what is called *outline processing*) and for excursions to related subjects, as well as browsing and search facilities.

Unfolding/folding can be used for quick navigation through a table of contents. In a cookbook, clicking on the first item of the following list

- MEATS
- FISH AND SHELLFISH
- VEGETABLES

will cause this line to "open up", so that the screen displays

- MEATS
  * *Beef*
  * *Pork*
  * *Veal*
  * *Lamb*
  * *Poultry*
- FISH AND SHELLFISH
- VEGETABLES

Repeating this process on a well-organised table can quickly lead to any desired destination.

A beginning chef might not know the meaning of some technical term, say "sauté". Clicking on such a term will display an explanation. An alternative possibility is to "macro expand" the sentence in which the term occurs. For example, "Sauté the slices of liver" is expanded into "Cook the slices of liver in a small amount of cooking fat or oil until brown and tender, using a skillet on medium fire and stirring well".

Note that the extent of navigation facilities need not be confined to the book itself; the book proper may be a coordinated collection of nodes in a larger web of active knowledge. This is of obvious relevance to mathematics (as well as many other fields of knowledge), and we come back to this in the next section.

2.2 Annotation facilities

As in a paper book, readers may want to highlight portions, or to jot down remarks, add references, and so on. Unlike a paper book, these are reversible actions, so that the reader need not feel the anxiety of possibly spoiling the book. Possible annotations in a cookbook could be a suggestion which wine to serve with a dish, or marks for excellent or disappointing dishes. Cooks can also add their own favorite recipes to the cookbook, right where they belong in its organisation.

2.1 Tailoring to user need

Given the interests and needs of the reader, the book can present slightly, or even completely, different versions of its text. An interactive cookbook can give the measures in recipes in pints and ounces, or in liters and grams, according to user preference. It can also present the actual quantities needed for the expected number of people, where a paper cookbook leaves it to the chef to do a multiplication and division, the latter by a number you can never find in the book when you need it. Likewise, the number of calories (or joules) can be computed and presented. The recipes presented can be filtered on the availability of ingredients or the time available for cooking. The book can even turn into a specialised cookbook for certain dietary restrictions: a vegetarian cookbook by omitting meat-based recipes and doing some standard substitutions; or a cookbook trimmed down to low-calorie recipes for people who are minding their waist.

There are more exotic possibilities. Given a menu (a collection of recipes, one for each course) the book can present a merged and sorted shopping list. Also, it can merge the preparation steps of the recipes for the various dishes into one time axis and produce one combined list of instructions, so that a cook can work in parallel on several dishes without ever having to jump to and fro between different recipes.

2.4 Expert assistance

An interactive cookbook can also assist a cook more actively. While going through the actual preparation, it can highlight the next step to be done. When the reader clicks the DONE button, it highlights the following step. In addition, it can serve as a timer and tell our cook when to lower the heat, et cetera. If things do not go quite as smoothly as planned, it can rearrange the steps on the time axis in the culinarily most appropriate way. Finally, if something goes really wrong, it could offer expert advice on the best way to save the meal – with “Call 1-800-DIAL-A-PIZZA” as the last resort.

Expert assistance can also be helpful in composing a menu. In its simplest form, the user composes a tentative menu, whereupon the book reacts with an evaluation, using basic rules about the desirable variation in a menu as well as “common-sense” and user-supplied constraints. In a more advanced setup expert assistance can be a truly interactive affair, with the book also suggesting modifications, and using the user reaction to those to adjust the constraints.

This does not exhaust the possibilities; neither in general (for example, we have not mentioned setting bookmarks), nor those specific to cookbooks (for example, planning ahead to use parts of a preparation for a later day, or giving advice what to do with leftovers), and, in fact, the possibilities are limited more by our imagination than by technical constraints. The preceding exposition should be indicative, though, of the nature of the extra value that interactivity can add to a book.

3 Interactive mathematical books

A mathematical textbook is not a cookbook. Yet, the reader will have few problems in transforming many of the examples given above of how interactivity can be helpful to a mathematical context. Although the examples were made concrete by infusing them with subject-specific detail, they reflect more general principles of the kind of support that is helpful to human beings when approaching a complex task, in particular in an area that they have not yet mastered. Thus, it might seem that any mathematical aspects are only in the content of the book, and not in the medium, just as for traditional books.

Yet, there is something unique to the mathematical world: it is a formal world that does not of itself refer to the outside world. Any such references, as in mathematical physics, are externally imposed interpretations on symbols that by themselves have no meaning grounded in reality.

To learn to cook, the aspirant cook has to get hands-on experience and stir the pots and pans. No amount of reading, however interactive, can fully replace that experience (a virtual-reality simulation tool can profitably substitute for some of the real-world learning experience; for cooking, though, it would also have to cater for the sensation of taste). In mathematics, the "real" world is the mathematical world "in here": if some piece of mathematics can be done at all, it can – in principle – be done now and here, in the medium of the book. This opens unique possibilities for active reader involvement.

Let us examine in more detail the interactive potential that is of specific value (although generally not exclusively so) in a mathematical context. We follow the same grouping as in the previous section.

Navigation support

Half of mathematics consists of giving definitions, and the other half of applying definitions. So the number of defined notions that are used is consistently high, and, more than in any other subject, the precise formulation of the definition counts. Thus, the ability to look up the definition of a mathematical term or symbol at the click of a button is especially important. Defined notions usually have arguments. For example, "is continuous" takes a function as argument, as well as some arguments that are usually implicit, namely the topologies of the domain and codomain. It should be possible to view an unfolded application with its actual arguments in place – including the implicit ones, which may become important if the definition is unfolded further.

Often not all mathematical knowledge used will be contained in the current interactive book. Thus, next to references to traditional texts, there is a clear value in allowing links to other interactive documents, mathematical encyclopedias in electronic form, etc., which will increase as more mathematical texts become available on-line. Although the specimen interactive book we intend to produce will not have external hyperlinks, our plan is to provide for the possibility, for example links based on the HTTP protocol of the World-Wide Web (see Berners-Lee et al. 1992), and experiment with them (see Fig. 1).

When studying a large theory it is very easy to get lost. In such cases a reader can be greatly helped when provided with an easily navigable presentation of the mathematical structure of that particular theory, which is basically a graph containing the definitions, lemmas, and theorems and their interrelations. This graph can be automatically extracted from the text if all hyperlinks are in place, and properly annotated as to their nature. Obviously, such a facility is not only useful for not getting lost; it may also be very helpful for authors, or for researchers who want to generalise a theory to a related area of mathematics.

Annotation facilities

For mathematics it is important that in an electronic book there is no reason for reader annotations to be limited by the size of the margin. Likewise, they need not be confined by the limitations of a purely textual rendering. Of the many possible kinds of annotations,

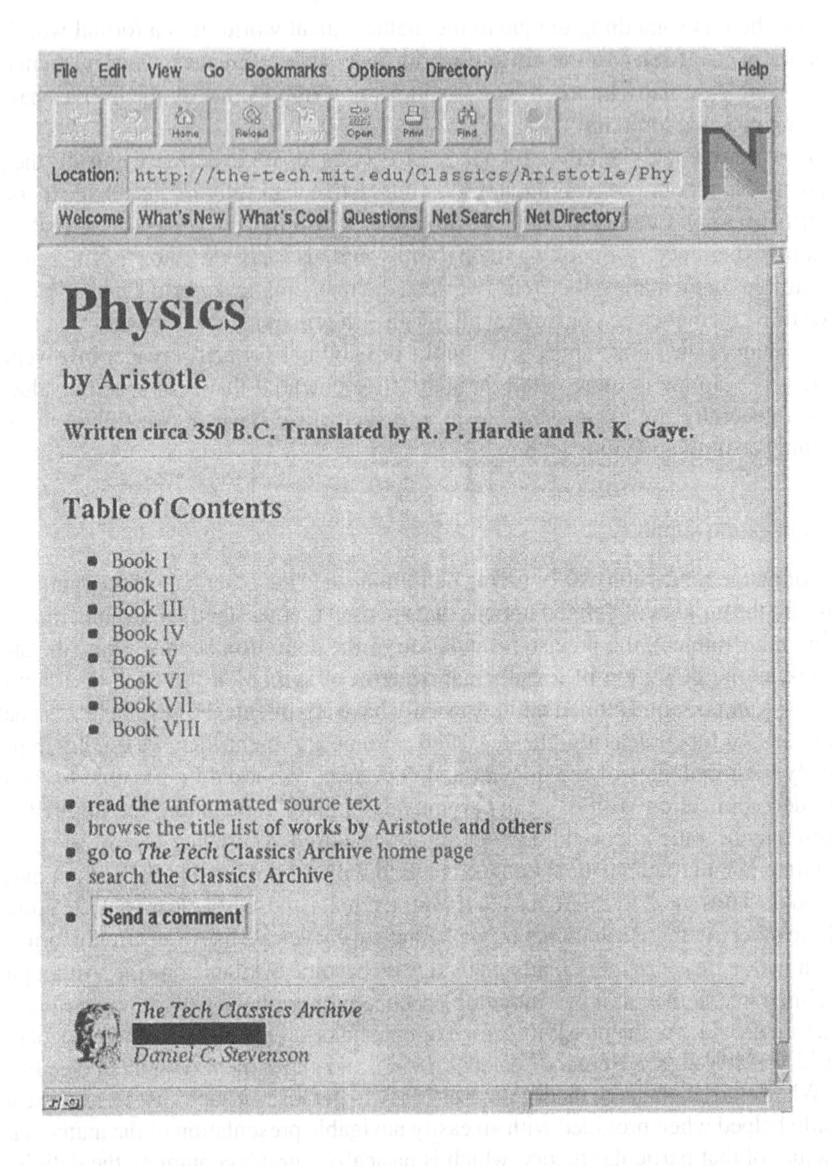

Fig. 1. A WWW page that could serve as an external hyperlink from an encyclopedia entry on Aristotle

we think those with mathematical content are the most important. Whenever the reader can ask the book to perform certain computations on mathematical objects contained in the original text of the book (say, compute the dimension of a given Lie algebra), this should be equally possible on mathematical objects occurring in annotations added by the reader. Thus, annotations are not second-class citizens, but must have recourse to the full spectrum of interactive capabilities.

Tailoring to user need

The book will allow readers to change the mathematical notations used to preferred notations, for example because of familiarity. The book will then also accept these notations for user-supplied input.

For advanced readers the book might hide from view elementary sections or expositions, give only informal and intuitive versions of the proofs, and skip elementary or add more advanced examples. It will also be possible for a reader to ask for a view on the book tuned to a specialised subject treated in it (for example, free Lie algebras), in which case it only presents those parts that are relevant to this subject.

More generally, there may be several linear routes through an essentially non-linear document, and the best choice may be determined by the interest and expertise of the reader. For example, a reader conversant with the topic of a book, consulting it as a reference on a specific issue, has other needs than the reader who is gradually learning about an entirely new subject, and this difference in needs can be reflected in a different presentation of the material. It is important here that the choices are under the control of the reader, and that locally other choices can be made than globally, for example for the professional reader who wants to take a quick "refresher course" in some specific part of a generally familiar area. The concept of routes and its implications for interactive books is worked out in more detail in Weber (1994).

Expert assistance

For interactive mathematical books, the recent and rapid development of high-quality computer algebra packages is of paramount importance. Provided the infrastructure needed to connect the two is present, a computer algebra package can bring life to the examples and exercises given in the book. Instead of one static example with predefined data, we can have *active examples*: the reader can change the data and immediately see the effect on the results. One might think of active examples where the reader has to choose, say, a simple Lie algebra and the book interactively computes and fills in all kinds of related expressions, for instance dimension, solvability, and so on.

In exercises, the book will provide hints and access to the computing machinery needed; tedious calculations can be left to a computer algebra engine, and the reader can concentrate on understanding the essence of the exercise. Furthermore, the book can (for some exercises) check the reader-supplied answer by simply computing it and, if the given answer is wrong, present a related exercise, which is recomputed for the given context.

Readers can use the book as a platform to perform their own mathematics within the context of the book. They will be able to insert solutions to exercises, and to invoke algorithms from computer algebra systems and other external applications transparently, meaning that the book takes care of the translation between the application's input and output format and the book's internal format.

The strict mathematical setup of the interactive book lends itself as a good test ground for further interactions with external mathematically relevant applications. Prominent examples of powerful tools of this kind come from logic, where semi-automated theorem provers and proof checkers have become available that are (on the verge of being) sufficiently sophisticated to be of help to mathematicians.

It is this promising integration of mathematical textbooks with recent developments in computer science, and especially computer algebra, that instigated us to start the ACELA project.

By taking Lie algebras as the subject matter for our first prototype, we hope to establish convincing evidence that an interactive book offers the reader a more active involvement than is possible with a paper book, a better guided tour than is provided by a computer algebra package, and a more rewarding experience than obtained by using these two together but separately, as independent entities.

4 Architecture

In designing and implementing the prototype system, we can distinguish between a subject-independent part – the kernel system – and a subject-dependent part – in our case, mathematics, and more specifically, Lie algebras.

The kernel system needs content, like a record player needs a record; in this case, the content is the text of the book. For a proper separation of concerns, we wish to consider the "truly" subject-dependent part as belonging to the content. So the content of a book can consist, next to a text, also of algorithmic entities: author-supplied *methods* to work on the types of objects occurring in the text. When the book "record" is loaded into the system "player", these methods are loaded as well. Part of the project is indeed devoted to creating powerful implementations of various algorithms for Lie algebras.

Still, we must take the future mathematical content into account right from the start, also for the design of the kernel system. The nontrivial demands posed by the requirement that mathematics can be brought to life would be virtually impossible to meet by afterthought modifications.

A simple example of this is the user control over the mathematical notation. This requires a strong logical separation of content structure and visual presentation, something that is missing in almost every mathematical-document preparation system (for a recent system that allows such a separation, see Backhouse and Verhoeven 1995). For example, although the *presentation* structure of $f(x)$ is `juxtapose(identifier ('f'), parenthesis(identifier('x')))`, its *content* structure is `apply (function=identifier('f'), argument=identifier('x'))`. If content structure and presentation structure are not kept separate in the software ab initio, it is practically impossible to pry them apart later. Such a separation is only possible if it is catered for in the architecture itself.

More generally, the desire to achieve transparent interoperability with external engines requires a flexible juggling with various representations for the same (mathematical) object. The actual representation rules are, of course, part of the (algorithmic) content of the book, but it would be awkward, to say the least, if the infrastructure for handling "Protean" representations and interposing the right transformations is not provided by the kernel system.

4.1 Interoperability with external engines

For the direct purpose of having a prototype system, we could resort to an architecturally less pleasing design and create an ad hoc interface to a Lie-algebraic engine. However, our plan is to approach the interoperability issues in a more general way: a further goal of the ACELA project is to establish a test ground for the transparent integration of computer algebra, document processors, and proof checkers, using a variety of external engines.

Our intentions with regard to proof checkers are twofold. First, we want to explore to what extent the mathematics presented in the interactive book can be formalised so as to be understandable to a proof checker such as Lego (Luo and Pollack 1992). Second, as logic-based proof checkers like Lego are notoriously inefficient in dealing with equational reasoning, the task of checking straightforward computations (both arithmetic and by symbolic calculus) had better be checked by a package more suitable for this task: a computer algebra system.

As an experiment, a definition of Lie algebra and a check that the commutator for an associative algebra defines a Lie algebra have been formalised, and checked by a combination of Lego and Reduce. Although, for this exercise, the interoperation has been programmed ad hoc, the experiment has shown that a nontrivial automated synergy between logic-based proof checkers and symbolic computation is feasible, as well as provided insight into the requirements for a more generic approach.

The book will provide the infrastructure for the interaction between, e.g., a proof checker and a computer algebra system. In order that a proof checker can ask the book to evaluate an equality involving symbolic computations, which the book can redirect transparently to a computer algebra engine, a connection with a computer algebra engine must be established. The representation that this engine requires for input is in general quite different from the proof checker's query: representations vary from engine to engine. From the point of view of the book proper, the editor can further be seen as yet another external engine. As the interactive book needs to communicate with all these applications, it needs to be able to convert the internal representation of an object to the various external representations and vice versa. This requirement poses several demands on the internal representation of a mathematical object. One key concept here is that of *application descriptions*: information relating the functionality of an application to the interpretational capabilities of the book, in particular giving the (co)domains and mapping rules for different operations as well as rules for maintaining a shadow copy of application-bound state information. Another key concept is that of *context*: a formalised description of the active set of mathematical knowledge in which the information is to be interpreted. [The information that in the present context f is a function and k a scalar makes it possible, for example, to recognise user input $f(n+1)$ as a function application and $k(n+1)$ as a formula with an implicit multiplication.] These concepts are worked out further in van Leeuwen (1994).

For our setup we need to use a protocol and exchange format for the exchange of mathematical data. While developing it, we will stay in correspondence with other groups, such as the Open Math project (which is developing a data exchange protocol to be added to Maple). Also of relevance is the SGML data type definition developed for the Euromath editor (von Sydow 1992).

4.2 User interaction

The infrastructure provided by the book for the interaction with external engines will also be used for the interaction with the user. Thus, the user can use familiar presentation-oriented notation while entering formulae, which are translated by the book into content-oriented structures.

We want the reader of the book to get into it with a minimum learning effort. The general experience is that users of interactive systems do not read the manual first. It is better to acknowledge this fact when designing the interaction. We hope, in fact, to design it in such a way that the reader does not need a manual to use the system, but can do with a small "try-me" interactive note on how to get started, together with adequate on-line help facilities. For the latter, we can use the basic navigation-support technology, provided that the user can do basic navigation without needing help for that. Providing access to the help facility by suggestively labelled buttons for hierarchical browsing, supplemented by an "enter-topic" query field, should do the job. More advanced navigation support should then be among the easily found help items.

For the more significant interactive potential, the expert assistance, it is helpful to the user if there is a uniform interaction paradigm. It may or may not be possible to obtain, through the book, *direct* access to some external engine, i.e., where the access is not mediated by the book. That depends on the contents of the book, authors can always create a book in which such direct access is possible, and if the topic of the book is, say, "Using Maple", direct access to Maple will be completely appropriate. However, exceptional cases apart, the general philosophy is that *all access to external engines is mediated transparently by the book*.

"Transparently" means here that the user interface keeps the same "look and feel", whatever the engine used. To achieve this we use the interaction paradigm developed in the Views project as described in Meertens and Pemberton (1992), specifically with the aim of creating a unified interaction model for the interoperability of varying applications (see Fig. 2). The conceptual model underlying the interaction paradigm of Views is that

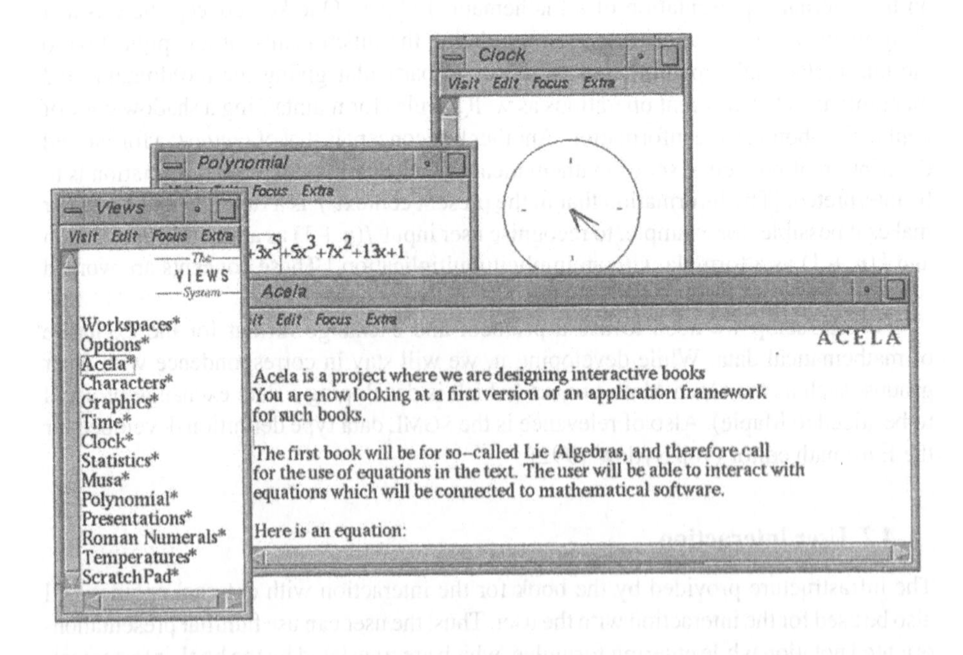

Fig. 2. Sample Views session

all interaction is achieved by the user by editing documents. These documents may be structured, and contain mathematical objects. The notion of editing comprises ways to enter and change text, as well as facilities for moving around and searching, in short, navigation.

In the Views interaction paradigm, performing some "symbolic" operation on a formula (like `evaluate` or `simplify`) is often best modelled as an edit operation that is performed in situ. By selecting a subformula, the operation can be restricted to only that subformula. The semantics of such an operation may depend on the kind of formula on which it is performed. Moreover, just as for a `print` operation, various options for a complicated operation can be set by the user in a "preferences sheet" for that operation. Ultimately, which method is invoked given all this, is determined by the author of the book. While the operation – which may possibly take a long time – is not completed, the reader will not be locked in but be able to continue with something else. The result of a still uncompleted operation is temporarily represented by a *promise*: a special object that can be copied, and all of whose instances will be replaced by the result upon completion.

An essential part of the Views interaction paradigm is that the semantics of the system as it presents itself to the user is based on defined logical relationships between various documents. If one is modified by the user, the system recomputes and updates the other. A basic example of this is a spreadsheet. A more involved example can be the relationship between a query on a database and the result of that query. By editing the query, the result is automatically updated. The same holds when the database is modified. Exercises and active examples can easily be created with this mechanism.

Similarly, the visual presentation of formulae is determined by one or more "notation" sheets, which are style sheets for the type or sub-types of "formula". By editing these sheets, the user can exercise control over the notation.

Since all users are familiar with the basic edit paradigm and edit mechanisms, getting to work with the intended interaction paradigm needs hardly any learning.

4.3 Authoring

The prototype system is not only intended to be for readers; the same system will be used by authors. Thus, there will be no distinction between a "reading" system and an "authoring" system. Obviously, authors have to be able to read what they have just written, and interact with it just like prospective readers could do, and readers, while creating annotations, become authors. The only hard difference is that, presumably, arbitrary readers cannot *change* the content of the book they are reading, but only change its *appearance* and further add annotations, while an author must be able to change the contents (what a reader can and cannot change in the book is ultimately controlled by the author, but in general readers will be better served if modifiability of the content is constrained to a few wisely chosen exceptions). However, we envisage the addition of a number of tools that, although accessible to readers, are primarily intended as facilities for authors creating a complete book text. For example, a constraint on a route through the text is that the sections may only depend on previous sections, and authors will be served with tools for managing routes that keep track of such constraints.

As for reading, the interaction paradigm for authoring will be that of editing text. The additional tools will facilitate structuring and restructuring, creating hyperlinks as well as the dependencies that are used for setting up routes, and interactively plugging in interfaces to external engines, together with monitoring and debugging aids.

For technical reasons, the first volume, on Lie algebras, will not entirely be built in this way: we cannot afford to wait until the prototype system is stable before starting to create the contents of the book. In addition, in designing the authoring facilities, we hope to learn from the experience obtained by the off-line creation of the text, in particular what need for facilities was felt.

5 The content of the book

5.1 The text

The actual topic of the first volume, that is, the specimen we shall produce as an interactive book, will be a new text on Lie algebras. Our minimum programme in the production of the necessary text is to cover the basic material on finite-dimensional Lie algebras, such as can be found in the well-known book, *Introduction to Lie Algebras and Representation Theory*, by Humphreys (1974).

Our goals are to produce a text that

- has the capabilities to serve as an introduction to the field;
- can serve as a platform and reference for a suite of Lie algebra algorithms;
- represents the state of the art in (algorithms concerning) semi-simple Lie algebras in characteristic 0, their (finite-dimensional) representation theory, and free Lie algebras;
- provides an infrastructure and experimental material for further experiments in interactive capabilities and in interoperability.

Humphreys has kindly agreed to serve as an advisor on the text. At present over 270 pages have been written (in a LaTeX-style markup language, with macros to indicate such things as hypertext anchors and links).

5.2 Algorithms

We intend to design and implement new Lie-algebraic algorithms. The documentation on the algorithms used should be quite extensive. In fact, we will strive for the following facilities:

- the possibility to invoke algorithms from within the interactive book;
- access to informal descriptions of the algorithms (in a Web-like pseudo-code form),
- access to fully readable well-documented source code.

The scope of the algorithms we envisage is roughly as follows:

- determining the structure of finite-dimensional Lie algebras;
- calculations with respect to representations of finite-dimensional Lie algebras;
- calculations with respect to free Lie algebras;

- related discrete-mathematical algorithms, such as representations of Weyl groups, Young tableaux and Zelevinsky pictures.

We intend to profit from the experience obtained in building LiE (van Leeuwen et al. 1992) and the Lie superalgebra package written by P. Gragert and M. Roelofs in Reduce.

A first implementation of algorithms for the structure determination of finite-dimensional Lie algebras has been written by de Graaf (1997) in GAP.

A first prototype of a discrete-mathematical algorithm was written by M. van Leeuwen and J. Faux. It represents the mathematical objects "Young tableau" and "Zelevinsky picture" in an attractive pictorial way on the screen, with possibilities to visually invoke algorithms like Schützenberger's "glissement" and the Robinson–Schensted correspondence.

6 Related work

Some close relatives (both in ideas and in scope) to our plans are the (concept of) Mathematica notebooks (Fultz and Grohens 1993) and the Euromath system (von Sydow 1992).

Both are based on proprietary commercial systems, Mathematica and the Grif editor (Quint and Vatton 1986), respectively. Since the source code of these systems is not publicly accessible, they are unsuitable as a vehicle for our research plans. Nevertheless, as an experiment, we are trying to set up a sample from our text on these systems to get a clearer picture of what hard limitations are encountered.

As to the Mathematica notebooks, these can be seen as a shell around Mathematica, which is not only a computer algebra engine, but also a document processor and a programming language in one.

In many respects Mathematica is a fine product, and for many subjects it is an easy tool for building an interactive textbook (see Fig. 3). However, it has several shortcomings that make the user interface of such a book more awkward than is desirable.

The internal representation of mathematical objects is dictated by Mathematica. It does not offer control over linking the interpretation of various symbols to the context in which they occur, and thus makes it impossible to achieve the interoperability we aim at. For example, in "x^2", the assumption is that "x" is a number, and not an element of some other algebra with a multiplication operator – like word concatenation in formal language theory.

The text of a Mathematica notebook is static, and it is impossible to tailor it to user need. Apart from an outlining facility, there is no hypertextual support. The activity comes from the reader talking directly to Mathematica. Thus, there is no hope of using the Views user-interface paradigm for an interactive book based on Mathematica.

Mathematica has also (somewhat rudimentary) facilities for communicating with external programs – all conversion has to be done on the other side.

The Euromath system is basically a kernel system consisting of a versatile generic structure editor, together with appropriate definitions of mathematical structures. The kernel system itself has, as in our envisaged architecture, no specific mathematical knowledge. There is a separation between logical structure and presentation structure, and the presentation rules are specifiable for each object type. The editor

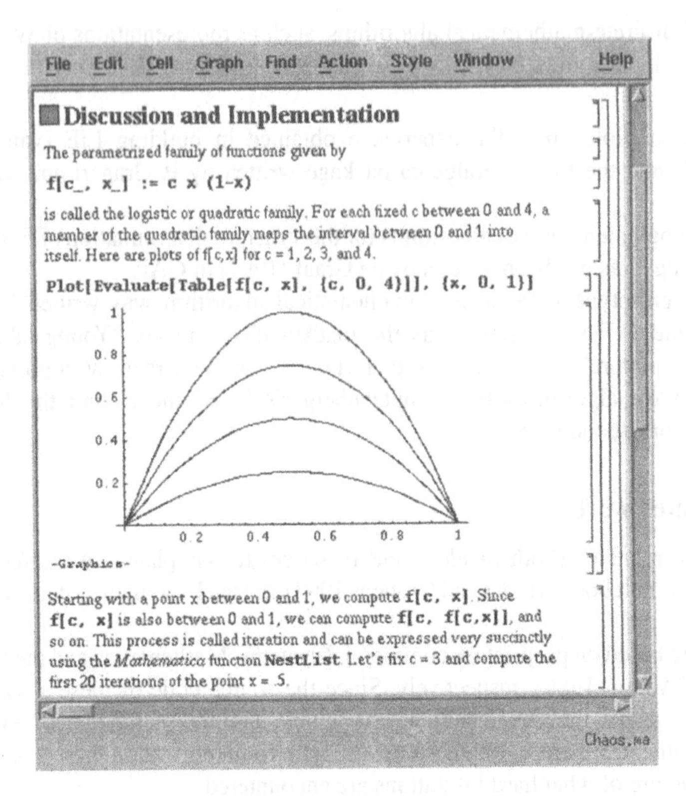

Fig. 3. Mathematica notebook

can handle most of conventional mathematical presentation, both for input and for output.

The set of edit actions is extendible, and the semantics of these actions can be an arbitrary piece of code, as described by Quint et al. in their contribution to this volume. So the Views interaction paradigm would seem consistent with this architecture, and to a certain extent it is; in fact, this possibility is an addition to the original Grif editor that was inspired by the Views paradigm.

However, unlike what we aim at and consider indispensable, the presentation rules can only be changed *off line*; during a session (a run of the system), they have to remain fixed. The concept of routes depends on the interactive computation and activation of a new presentation, and can therefore not be modelled (except by computing a complete new text).

As in Mathematica, there is no support for interpreting formulae in context. For interfacing with external engines, there is a translation facility, which, however, works only in one direction: outward. Thus, text can be shipped to, e.g., LaTeX; but there is no convenient way to receive results from computer algebra engines.

Most of the limitations mentioned above for Euromath and Mathematica are addressed in CAS/PI (Kajler 1992), an experimental system whose central notion is the interoperability of various engines: the internal representation of text and formulas is

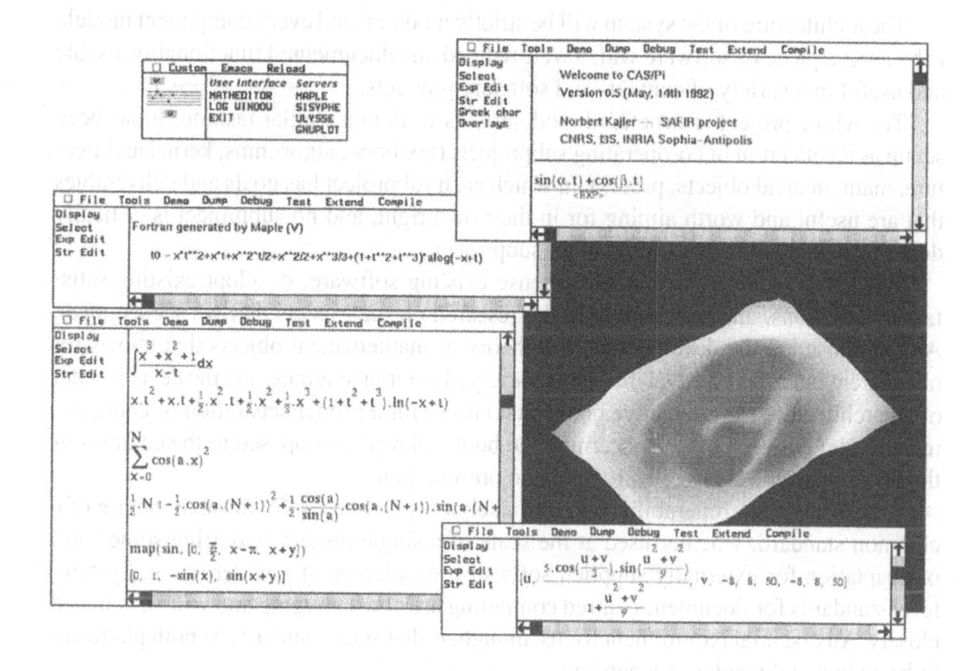

Fig. 4. Sample CAS/PI session

independent of the display views and the external views, and the necessary translations are handled transparently. CAS/PI offers the user a graphical interface to a formula editor, in which it is possible to mix text and expressions (but not graphs), and to select subexpressions via the mouse and request in situ evaluation (see Fig. 4). The graphical notations, as well as the menus and mouse-based edit actions, are extendible at run-time. Like Mathematica, CAS/PI offers no hypertextual support.

7 Conclusion

The aims and plans we have set forth in this paper are ambitious, and having some priorities among them is not a luxury. One priority is that we strive for getting the things that we do right, rather than getting necessarily everything done. As we see it, the architectural decisions made now will have a strong impact on the long-term potential of the system, and ill-chosen decisions will eventually emerge as unsurmountable limitations, but given the right architectural setup, further enhancements in functionality will always be possible.

None of the individual aims we have is by itself unique, in the sense that some product exists that achieves it. At the same time, in its current incarnation each existing product has limitations that make it unsuitable as a platform for realising the interactive book we have in mind. The main contribution of our project in this context would be in showing how the various desiderata can be combined gracefully.

The architecture of the system will be strictly modular, and every component module is by itself a piece of software with a well-defined and documented functionality, usable and useful in a variety of contexts and software products.

The whole project is structured modularly as well, in a similar fashion. It has been set up as a collection of co-operating subprojects (textbook, algorithms, kernel architecture, mathematical objects, proofs), in which each subproject has goals and deliverables that are useful and worth aiming for in their own right, and no subproject is critically dependent on the results of some other subproject.

Whenever possible, we intend to reuse existing software, or adopt existing satisfactory solutions, and concentrate in the research on the aspects that we see as novel. Among those are the development of a theory of mathematical objects that allows for transparent interoperability with a large variety of external engines, and the development of an architecture for interactive books that allows for a proper separation of concerns, in particular between those concerning the book "player" as opposed to the contents of the book, as well as between structure and presentation.

Any form of interoperability can ultimately only be successful through the use of a common standard, whether used at the scale of a single project (requiring some form of adaptation for externally supplied software), or adopted at a wider range. System-level standards for document-centred computing are now emerging and will be watched closely. An essential requirement for us, though, is that such standards be both platform-independent and vendor-independent.

References

Backhouse, R., Verhoeven, R. (1995): Math∫pad: ergonomic document preparation. Department of Mathematics and Computing Science, Eindhoven University of Technology, Eindhoven, The Netherlands.

Berners-Lee, T. J., Cailliau, R., Groff, J.-F. (1992): The World-Wide Web. Comput. Networks ISDN Syst. 25: 454–459.

de Graaf, W. A. (1997): Algorithms for finite-dimensional Lie algebras. Ph.D. thesis, Eindhoven University of Technology, Eindhoven, The Netherlands.

Fultz, J., Grohens, J. (1993): Mathematica user's guide for the X front end. Wolfram Research, Inc., Champaign.

Humphreys, J. E. (1974): Introduction to Lie algebras and representation theory. Springer, New York, Berlin Heidelberg.

Kajler, N. (1992): CAS/PI: a portable and extensible interface for computer algebra systems. In: Wang, P. S. (ed.): Proceedings of the International Symposium on Symbolic and Algebraic Computation, Berkeley, California. Association for Computing Machinery, New York, pp. 376–386.

Luo, Z., Pollack, R. (1992): The LEGO proof development system: a user's manual. Report ECS-LFCS-92-211, Laboratory for Foundations of Computer Science, University of Edinburgh, Edinburgh, U.K.

Meertens, L., Pemberton, S. (1992): The ergonomics of computer interfaces: designing a system for human use. Report CS-R9258, Centrum voor Wiskunde en Informatica, Amsterdam, The Netherlands.

Quint, V., Vatton, I. (1986): GRIF: an interactive system for structured document manipulation. In: van Vliet, J. C. (ed.): Text processing and document manipulation: proceedings of the International Conference on Text Processing and Document Manipulation. Cambridge University Press, Cambridge, pp. 200–213.

van Leeuwen, A. (1994): Representation of mathematical objects in interactive books. ACELA project working document, Centrum voor Wiskunde en Informatica, Amsterdam, The Netherlands.

van Leeuwen, M. A. A., Cohen, A. M., Lisser, B. (1992): LiE, a package for Lie group computations. Computer Algebra Nederland, Amsterdam, The Netherlands.

von Sydow, B. (1992): The design of the Euromath system. Euromath Bull. 1/1: 39–48.

Weber, O. (1994): Routes through the maze. ACELA working document, Centrum voor Wiskunde en Informatica, Amsterdam, The Netherlands.

Active structured documents as user interfaces

Vincent Quint, Irène Vatton, and Jean Paoli

1 Introduction

Mathematicians manipulate complex abstract objects and expect some help from the computer in this task. A number of systems have been developed for that purpose. The early developments focused on methods and algorithms for numerical and symbolic computations, without paying too much attention to the user interface of systems using these algorithms. Other tools have been developed for helping computer users to prepare mathematical documents. This trend is illustrated by the famous TEX system (Knuth 1984) that most mathematicians use nowadays. Here again, the emphasis was put on the algorithms and on the quality of the result, but the language provided to the user is not very user-friendly, although very powerful.

These two categories of systems (numerical or symbolic computations and document preparation systems) have something to do with each other. The result of a computation is often included in a document, and conversely, a formula found in a document can be submitted to a computer algebra system. In addition, the user interface problem is quite the same in all these systems: a convenient way of entering and modifying formulae is needed. In fact the problem is not restricted to formulae, but it also concerns other graphical representations used in mathematics, such as various types of graphs and tables.

As there are some similarities and interactions between symbolic algebra systems and document manipulation systems, it is tempting to address the user interface problems of both types of systems with the same approach and to solve them with the same techniques. This is what we do in this paper. We propose to use advanced document concepts, such as logical structuring and active documents, for making the user interface component of computer algebra systems and other types of interactive systems which are used when doing mathematics.

The next section briefly presents the toolkits that are currently proposed for building the graphical user interface of interactive systems; their strength and weakness are analyzed, specifically in the context of mathematical applications. Section 3 presents some advanced concepts used in document manipulation systems and focuses on their application to user interfaces. An implementation of these concepts, the Grif system, is described in Sect. 4. Section 5 gives several examples of applications built with this system. Related work is presented in Sect. 6.

2 Limitations of graphical user interface toolkits

A number of formalisms are used for representing the abstractions manipulated by mathematicians and physicists. They range from different kinds of mathematical expressions to conventional documents, including many types of structured graphics and tables. Most of them take a complex graphical form and need a sophisticated layout, with various styles of characters and very precise positioning. Because of the importance of these formalisms, computer algebra systems need efficient methods and tools for implementing a graphical user interface that allows users to interact with such formalisms.

Currently, a number of graphical user interface toolkits are proposed, such as the Macintosh tool box (Apple Computer, Inc. 1991), OS/2 (IBM 1991), OSF/MOTIF (Open Software Foundation 1992), and Microsoft Windows (Microsoft 1992). From our perspective, they can be considered as very close to each other. All have the same limitations when considering the kind of manipulations a mathematician does.

Most of the available widgets are dedicated to the dialogue between a user and an application, by means of buttons, menus, dialogue boxes, etc. These widgets are obviously useful, but the most complex part of a mathematical user interface is not at that level.

They offer only low-level functions for manipulating text: it is often difficult to handle mathematical symbols and multi-font formatted text in different sizes.

Complex pictures including formatted text, tables, equations, graphics, etc. are outside the scope of such toolkits; their layout capabilities are too poor. It is always possible for an application to compute the size and position of each character, symbol, or line, and to ask the toolkit to display it on the screen, but this task is very cumbersome. It should be accomplished by the toolkit, in the same way it handles simple text.

Toolkits do not provide any mechanism for handling interaction on formatted pictures. If a formula is displayed by the application as a collection of elementary characters and symbols, the user cannot select a given part of the formula without some computation made by the application. If the user inserts or deletes a character within a formula, the application must compute the new position of other characters and symbols; it must evaluate which ones move and it must redisplay them. Toolkits often offer these operations for character strings, but they do not provide anything equivalent for handling mathematical objects.

Due to these limitations, development of applications that handle mathematical formulae and graphics with a direct manipulation style of interface is not made easier by these toolkits. For manipulating complex pictures interactively, tools offering higher-level functions are needed. At least, the operations that available toolkits provide for manipulating simple character strings should also be provided for manipulating more complex objects.

3 Active and structured documents

Document editors are worth being considered for offering the high-level functions needed. They can format complex pictures (Quint and Vatton 1987) and they allow users to interact with these pictures. Many of them handle tables, equations, and structured graphics in addition to multi-font text. Most computer users are familiar with these applications and can use them easily; hence the idea of using a document editor as a user interface component for an application that handles complex objects.

Simple word processors are obviously not sufficient. Systems that can handle more complex documents are needed. Most of these systems are based on a high-level document model that considers a document as a logical structure. In addition, the recent evolution of these systems towards active document environments seems very promising. In this section, we present the concepts of structured documents and active documents.

3.1 Structured documents

The structured-document approach (André et al. 1989) considers a document as a logical structure, not simply as a collection of characters or graphics printed on paper or displayed on a screen. Of course, structured documents can be printed or displayed, but for allowing various applications to process them in different ways, they are represented in an abstract way. Formatting is only one of these possible applications, which produces a graphical representation of a document.

With that approach, a document is a hierarchical structure (called an *abstract tree*) that gathers typed elements, such as chapters, sections, paragraphs, lists, figures, notes, etc. (see Fig. 1). Elements may have attributes that add semantics to the element type. Attributes, element types, and their allowed structural relationships in the abstract tree are defined by *generic logical structures* that can be compared to grammars. Each generic logical structure defines a document type. This approach is now well accepted. It is the basis of such international standards as SGML (I.S.O. 1986) and ODA (I.S.O. 1989) and more and more systems are adapted to (or developed for) these standards.

The structured-document approach is used for representing not only the textual part of documents, but also components such as equations, tables, or structured graphics, with the advantage of a uniform representation of all parts of a document. Concerning equations, the structure can be similar to the one used in computer algebra packages. It can even be made to match the one used in a given package, as the generic logical structure

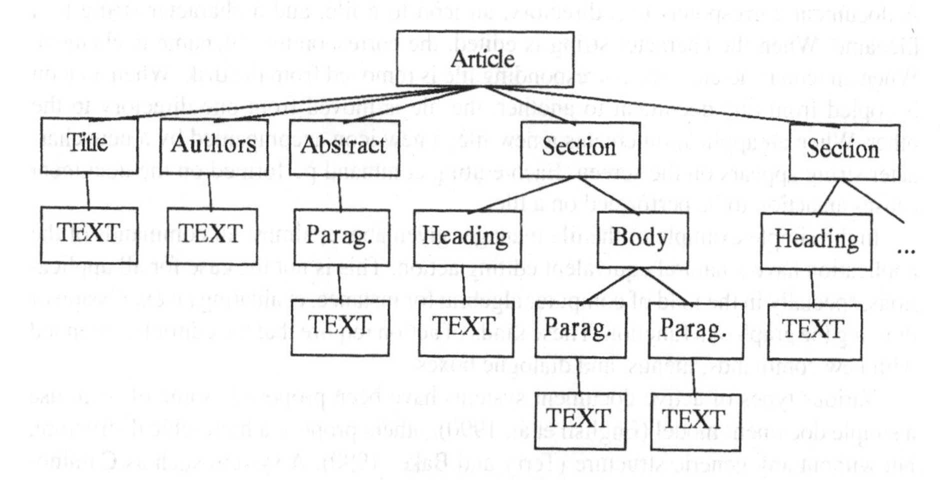

Fig. 1. An abstract tree

(also called Document Type Definition (DTD) in SGML) allows the specification of such a structure. This feature, of course, facilitates the connection of a structured-document editor with a computer algebra system.

In most document editors based on this model, the user does not see the abstract structure on the screen, but, instead, a formatted representation of the document. The editor produces that picture from the abstract structure, by applying *presentation rules* which specify how each type of element should be formatted, according to its structural position and its attributes. As these presentation rules are not part of the document, they can be changed without any modification to the document itself, thus allowing the user to see (or print) the same document in different ways. This operation gives flexibility to the user interface, as it allows each user to see things according to her skills and taste or to the type of activity she is performing on the document.

3.2 Active documents

Although structured-document editors provide a number of interesting features for making user interfaces, there remains an important problem: document editors are not as flexible and reusable as user interface toolkits are. They must be open-ended in such a way they can be integrated in a variety of applications.

This can be achieved with the concept of an active document. An active document is a document that acts on its computing environment or transforms itself when it is manipulated by a user through an editor. This approach seems well suited to user interfaces (Bier and Goodisman 1990). For controlling an application, a user edits a document whose elements represent the objects handled by the application. When editing actions are performed by the user on a document element, the application performs some equivalent functions on the corresponding object. Conversely, when the internal objects are changed by the application, the document is updated, in such a way that it reflects the current state of internal objects.

The desktop metaphor can be considered as an example of an active document used as a user interface. In the case of a file manager, for instance, the document displayed on the screen is simply a collection of icons accompanied by short character strings. A document corresponds to a directory, an icon to a file, and a character string to a filename. When the character string is edited, the corresponding filename is changed. When an icon is deleted, the corresponding file is removed from the disk. When an icon is copied from one document to another, the file is moved from one directory to the other. When an application creates a new file, a new icon accompanied by a new character string appears on the screen. Each editing command performed on the document causes an action to be performed on a file.

In the simple example of the file manager given above, almost all commands of the application have a natural equivalent editing action. This is not the case for all applications, specially in the field of computer algebra; for instance, evaluating an expression or drawing the graph of a function. These kinds of action require that the editor be extended with new commands, menus, and dialogue boxes.

Various types of active document systems have been proposed. Some of them use a simple document model (English et al. 1990), others propose a hierarchical structure, but without any generic structure (Terry and Baker 1990). A system such as Camino-Real has specifically been designed for handling equations as active documents (Arnon et al. 1988). But systems based on structured documents are especially interesting in the field of computer algebra applications, as they combine the advantages of a rich logical structure and the power of active documents. The next section describes such a system.

4 The Grif editing toolkit

We have been working for a long time on structured documents and associated tools. An editor for structured documents, called Grif (Quint and Vatton 1986), has been developed and more recently an editing toolkit (Quint and Vatton 1994) has been derived from that editor.[1] Both the editor and the editing toolkit are based on the concepts presented above. Both use generic logical structures (or SGML DTDs). The editor helps users to produce documents (abstract trees) that are consistent with a given generic logical structure. The toolkit helps programs to produce documents in the same way. Both guarantee that any document they produce is structurally correct regarding the generic logical structure. This is a great advantage, as any application that works on these documents knows how the document is organized, just by consulting its generic logical structure; it can directly access any part that is of interest to it.

4.1 Document editor

In the Grif editor, the generic logical structures are used to specify various types of documents and objects, such as books, articles, letters, reports, but also tables, equations, and various kinds of structured graphics. (In fact, several generic logical structures (GLS) may be involved in a single document, for instance when the document, itself defined by a GLS, contains tables defined by another GLS and different types of graphics defined by different GLS.)

Presentation models

Presentation rules are available. The editor applies them to the abstract-tree elements to produce the graphical appearance of the document. Presentation rules are expressed in a declarative language called P, that allows one to specify such properties as font, color, line length, indentation, justification, etc. These rules also specify the relative dimensions and position of the elements according to their type and their structural position. Presentation rules are gathered in *presentation models*. A presentation model contains all those rules needed for specifying the presentation properties of all types of elements and attributes defined in a generic logical structure.

The editor uses presentation models when it generates the graphical form of (parts of) documents displayed on the screen. This formatting process is obviously interactive and incremental, as each elementary change made to a document is immediately reflected on the screen. Simply by changing the presentation model, it is possible to completely

1 The Grif technology was developed at INRIA (Institut National de Recherche en Informatique et en Automatique) and CNRS (Centre National de la Recherche Scientifique) and is industrialized and commercialized by Grif S. A. Set up in 1991, Grif S. A. developed and brought to market the first range of user-friendly WYSIWYG SGML authoring tools.

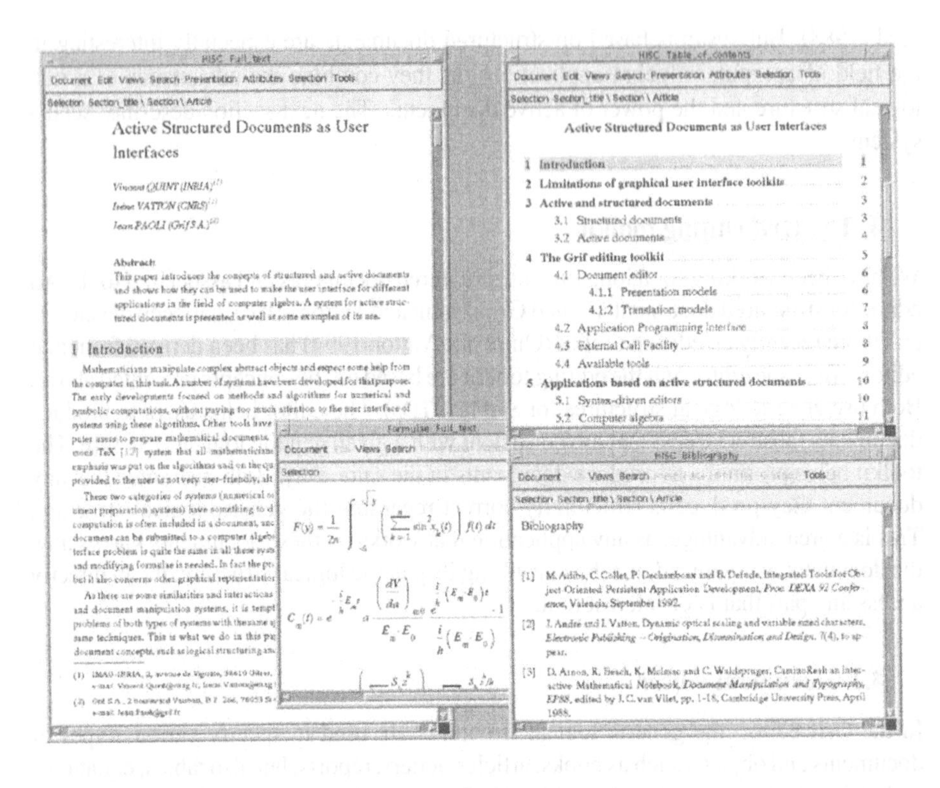

Fig. 2. Several views of a document

change the graphical appearance of the document on the screen, and this is possible at any time.

Presentation models also define different *views* which present the same abstract tree with different graphical appearances in different windows at the same time (see Fig. 2). For instance, the table of contents of a report is a view that shows only the elements of type SectionHeading and does not use the same fonts and layout properties as the main view. These views show a graphical representation of documents that is very similar to a printed document: the abstract tree is not directly visible to the user, who interacts through the views displayed on the screen. Any view can be used for any editing command, provided the element(s) concerned is (are) visible. As soon as any change is done, the editor automatically reflects it in all views, thus maintaining the consistency of the display.

Translation models

As a document is represented in an abstract form, it is possible to derive different concrete representations of it. Presentation models allow generation of graphical representations of documents. There exist also *translation models*, which allow generation of textual representations of documents. A translation model gathers *translation rules* which indicate how each type of element or each attribute should be translated to represent the document in a given syntax. A program, called translator, is associated with the editor.

It interprets these rules and produces an external form of a document. This is used, for instance, for producing the LaTeX representation of a document or its SGML representation. Translation models can be used to translate a formula into the Maple language, or any other language of that kind.

Translation models are much simpler than presentation models. In order to allow high-quality displays to be produced, a presentation model must specify a number of graphical properties for each type of element and for each attribute defined in a generic structure. Due to an inheritance mechanism based on the logical structure, it is not necessary to explicitly specify all properties for each type and attribute, but a rich presentation model for a complex logical structure can be fairly large. Conversely, translation models are much simpler, as they have only to specify some (generally short) character strings to be generated when some types of elements or some attributes are encountered in a given context in the logical structure. Nevertheless, presentation models and translation models share the same principle of generating different types of data (graphical properties and character strings, respectively) depending on the logical structure. Therefore, their languages share many common features.

A structured-document editor such as the Grif system presented above is not enough for handling active documents. Specific mechanisms must be added to allow it to interact with other applications. It should be possible for the editor to call an application, and for the application to call the editor. Therefore, two different mechanisms have been added to the editor:

- an *application programming interface* (API) allows an application program to call editing functions,
- a mechanism called *external call facility* (ECF) allows the editor to trigger some activities in an application on some specific events.

4.2 Application programming interface

The API is a comprehensive set of editing functions that can be used for constructing applications which handle structured documents. It allows application programs to perform the same operations as a user working with the Grif editor.

Typically, the API provides an application with functions for creating the skeleton of a new document of a given type, for modifying the structure or the contents of existing documents, for extracting information from documents according to their structure, for opening and closing documents and views, etc. About 150 functions are available. It also gives access to the Grif user interface, and to some functions for extending this interface.

When using the API, applications have not to worry about formatting or displaying; they only have to handle the abstract tree and the contents of documents. Using the presentation models, the editor automatically takes care of all problems related to formatting and refreshing the pictures displayed on the screen, when changes are made in the abstract tree by an application program.

4.3 External call facility

The ECF is a mechanism that allows applications to control the user's actions. In the desktop metaphor example (Sect. 3.2), when the user modifies a filename, the application must check the validity of the new name, for avoiding an existing file to be replaced by mistake. The editing toolkit cannot check that by itself, as the generic logical structure cannot express this kind of constraint. Then the file management application needs to be notified by the editor every time a filename is entered or modified by the user. If the new name is invalid, the application may then display a message for the user and change the document through the API (e.g., by modifying the entered name).

The ECF mechanism is based on the abstract tree and its generic specification. It allows a developer to declare how and when the editing toolkit should give control to the application. This mechanism is similar to the callbacks provided by many user interface toolkits. The Grif editor generates *events* and executes application procedures (called *actions*) in certain conditions. For instance, an event occurs when the user selects an element, modifies a character string, associates an attribute with an element, saves a document, opens a view, creates or deletes an element.

Each event is associated with the object concerned by the corresponding editing operation: logical element, attribute, document, view. Each event can occur only on certain objects. For instance, events "Selection", "Creation", and "Deletion" concern only logical elements; event "Attribute association" concerns both an attribute and an element (the element with which the attribute is associated by the user); event "Saved" concerns only a document.

An important difference to callbacks provided by most user interface toolkits is the ability of the ECF to control event generation: events to be generated for a given application are defined in generic and contextual terms. For each type of document it manipulates, an application declares in what events, associated with what types of object, it is interested. It also indicates what actions must be executed when these events occur for these types of objects. As an example, the above mentioned file manager is interested in the event TextModified associated with elements of type FileName (assuming that this type is defined in the generic logical structure of the documents processed by the application). Then the Grif editor calls the desired action each time the user modifies any file name.

The Grif system offers a declarative language to specify event generation and associated actions. It automatically generates the skeleton of the procedures that implement the corresponding actions from this declaration. The programmer has only to write the code of these procedures.

4.4 Available tools

The Grif editor, the API, and the ECF presented above are available for several Unix, PC/Windows and Apple Macintosh. They are proposed by Grif S.A. in a range of products:

- Grif-SGML Editor is an editor for structured SGML documents designed for authors;
- Grif Application Builder creates SGML DTDs and presentation models;
- Grif Application Toolkit Environment (GATE) contains the API and the ECF mechanism.

5 Applications based on active structured documents

Various kinds of document-centered applications have been developed with the mechanisms presented above, in several fields. This section presents such applications in software engineering (syntax-driven editors for programming languages), computer algebra, document production (typing and formatting complex formulae), hypertext, data bases access, etc.

5.1 Syntax-driven editors

The syntactic structure of a program is very similar to the abstract tree that represents a structured document. The grammar of a programming language can be represented to a large extent as a generic logical structure, such as the one used by Grif for specifying document types. In addition, a presentation language like the one used in Grif can specify a graphical appearance that can be either textual or graphical, or a mixture of text and graphics. For these reasons, Grif has been used as a toolkit for building program editors: by writing only a generic logical structure and a presentation model, a developer can make the basis of an editor for a given programming language.

Obviously, the semantics of a program cannot be fully represented in a structured document. The application, which knows about the language semantics, must control editing commands, preventing the user from performing commands leading to incorrect programs. For example, it checks that each variable used is declared and it proposes to add its declaration if it is missing. The application can also offer specific commands depending on the language semantics. Two such editors have been built by using the Grif active-document mechanisms. An editor has been developed for the database-programming language Peplom (Adiba et al. 1992) and another for the real-time language Argos (Maraninchi 1991). The graphical possibilities of Grif have proven very useful in the case of Argos, as the language has both a textual syntax and a graphical syntax, which are simultaneously presented in separate views.

Both editors are purely structure-driven editors. They do not include any parser or compiler, as they know the programming-language grammar. This knowledge comes from the generic logical structure and from the application itself. This can be compared to the Grif SGML editor. Except when loading an existing document, this editor does not parse SGML syntax. It works only on a logical structure, driven by a generic structure and it produces the SGML syntax only when the document is saved in a file.

5.2 Computer algebra

In a computer algebra environment, the approach presented above has a number of advantages. It can be used to make the user interface of a computer algebra package in the following way.

A generic structure of mathematical expressions is first defined. It must be close to the structure of equations manipulated by the package, for making the interface between the editor and the package simple.

Then, a presentation model is associated with the generic structure to define the graphical aspect of formulae. The result is a formula editor that can be immediately tested and validated.

A translation model is specified for converting the structure produced by the editor into the syntax accepted by the package.

Finally, the editing events that should trigger some action in the package are specified and the code of these actions is written.

These four steps allow to use the Grif editor as a user interface component for a computer algebra package. Interaction with computer algebra applications is then simply perceived as modifying a mathematical document. The connection of the Euromath system (von Sydow 1992) to MuPAD (Fuchssteiner et al. 1993) illustrates this method; it should be noted that the mechanisms developed by Euromath are generic, i.e., it could be used for other computer algebra systems. The connection to MuPAD is the first instantiation of this generic mechanism.

The Euromath system is based on Grif. It is a WYSIWYG editor which manipulates structured documents. In these documents mathematical formulae belong to the Euromath DTD, a DTD designed to describe the structure of formulae, very much like TEX does. In the editor, formulae are represented in a structured way, following this DTD.

MuPAD is a modern computer algebra (CA) system, developed at the University of Paderborn, which supports full parallel processing capabilities. To connect it to the Euromath system, three main issues have been addressed: the connection itself, the data exchange format, and the user interface.

Connection management

The aim of the connection is to control the CA system and to make it evaluate expressions from within the editor.

The connection is based on a client/server architecture. It is connection-oriented, i.e., the link exists during the whole CA session. A limitation in the current implementation is that closing the connection implies terminating the CA session.

Data exchange

The connection implies bidirectional exchange of CA-specific data. Therefore, an exchange format has been defined between the two systems.

The documents manipulated in the editor are all based on SGML, according to the Euromath DTD for formulae.

It has been established (Harbo et al. 1993, von Sydow 1994) that one cannot expect to capture the complete semantics of an expression in a DTD aimed to permit user manipulation of formulae and their interactive display on screen. What is expressed in the Euromath DTD is only the syntactical structure, i.e., how subexpressions are related to one another.

One could have chosen to translate directly formulae written in this DTD to the expression language used by MuPAD. Direct translation would have represented the least amount of work and made it feasible to support specific features of the MuPAD system. Another solution has been adopted: an intermediate data-interchange format has been defined. Mathematical expressions represented in the editor are translated to this intermediate format which is then translated to the proprietary MuPAD data format. In the reverse direction, results from MuPAD are translated in this intermediate format and then translated in the Euromath DTD. This interchange format has been expressed in the SGML syntax, in another DTD which caries more semantics.

One cannot here use directly the Euromath DTD because this DTD is an intermediate step between a semantically oriented DTD and a display-oriented DTD. For example, the Euromath DTD defines a root construct but basic expressions, such as $ax + b$, are represented by a single character string. The second DTD gives more mathematical details about these basic expressions. One could have used the OpenMath protocol for communicating with MuPAD instead of this second DTD, but at that time, OpenMath was not yet defined.

A solution based on two DTDs ensures a clear definition of the borderline between the editor and the integrated CA system.

User interface

The use of the Euromath system provides the user with a formatted display of the manipulated formulae. The user constructs a formula with the editor, and submits it from within the editor to the CA system. The output returned by the CA system is presented to the user into the document in a well formatted form.

The user is able to re-use the output returned by the CA system. He/she can edit it again and submit it to the CA system. He/she can interrupt the CA system during computation and, when an expression has been sent off to the CA system, the user can continue to work with the editor, i.e., the Euromath system does not block while waiting for the CA system to respond.

5.3 LaTeX and structured editing

Editing formulae is a critical task for mathematicians. By construction, the Grif editor allows a structured manipulation of formulae, by using menus of mathematical constructs or their equivalent function keys. But it is often tedious to enter some parts of mathematical expressions with structure manipulation commands, and users want to type formulae very rapidly. This can be achieved by allowing them to input a mathematical expression as a simple character string.

This way of entering formulae has been included in the Euromath system. As almost all mathematicians know the LaTeX syntax or its variants, this facility has been based on LaTeX (Timoney 1994). It permits to turn any structured formula included in a document into its equivalent LaTeX representation. Conversely, it permits to turn any formula written in LaTeX into its representation in the Grif logical structure. A single key stroke, Meta-$, allows to switch back and forth. The LaTeX form is displayed instead of the formatted formula, at the same place, and the user manipulates it as a character string. This allows fairly complex restructuring to be performed in a simple way. It is also convenient for some users who are very comfortable with LaTeX to type new formulae very fast. When the user types Meta-$, the character string is parsed and the resulting formatted formula replaces the LaTeX form.

Parsing is performed by a C language program generated using the lex and yacc Unix tools. The parser reads a LaTeX string representing a formula such as `\frac{-a}{x^2}` and produces the equivalent logical structure in the abstract tree of the Grif document. In the example, it produces a fraction structure where the numerator is a single textual item while the denominator contains another structure, that of a "superscripted expression", with two further substructures. The construction of this structure and its integration into the document abstract tree are done through the API.

The reverse transformation, from the Grif abstract tree to LaTeX, is simply performed by the Grif translator, according to a specific translation model.

This way, when manipulating mathematical formulae, the user can freely mix interactive structured editing with text mode editing.

5.4 Interactive formatting of complex formulae

Another extension to the editor has been developed with the API and ECF for fine tuning the layout of formulae. Many mathematical symbols (surd, parentheses, braces, integrals, etc.) have to be sized according to the expression they enclose. With a structured representation of mathematical expressions and with presentation rules based on this structure, the editor can compute the right size of such symbols. But this is not always enough, as the position and size of symbols may depend on a number of constraints that cannot be expressed in the P language. (The P language has been designed for allowing the editor to efficiently compute the position and size of all components of a document in real time, while the user is typing. Therefore very complex constraints have not been considered.) For example, if two integral symbols are displayed on the same line of a mathematical expression, they should be displayed with exactly the same height, even if their integrands have slightly different heights; but if the integrands have very different heights, the integral signs must be different. Using the API and the ECF, an application can resize integrals (and other such symbols), on some specific events, for instance when changes are made in integrands (André 1994).

5.5 Other applications

Other applications have been developed with Grif active structured documents, in various fields.

A specific editor has been derived from the Grif system and tailored for the World-Wide Web. This Web editor (Paoli 1995, Quint 1995) allows its user to create and edit HTML documents and to manipulate in a user-friendly way the links that relate a document to other documents across the World-Wide Web (HTML is an SGML DTD that defines the structure and the syntax of the documents made available through the World-Wide Web). The specific Web functions have been developed with the API and the ECF.

In an illustrated parts list catalog (see Fig. 3), keying the content of a part reference automatically queries the database for the part description while keying the same content in the title of the document does nothing. A database menu is also available for the part reference to give the list of valid choices.

In a business application, when putting together a commercial offer, the action of keying in some text can incorporate pieces of documents such as a product description, bill of material or contractual items.

In an industrial application, a test technical manual linked with a database contains many images where sensitive transparent areas are defined. A sensitive area is a rectangle which is superimposed on a portion of an image and a mouse click on the area queries the database and loads the pertinent information which describes the portion of the image.

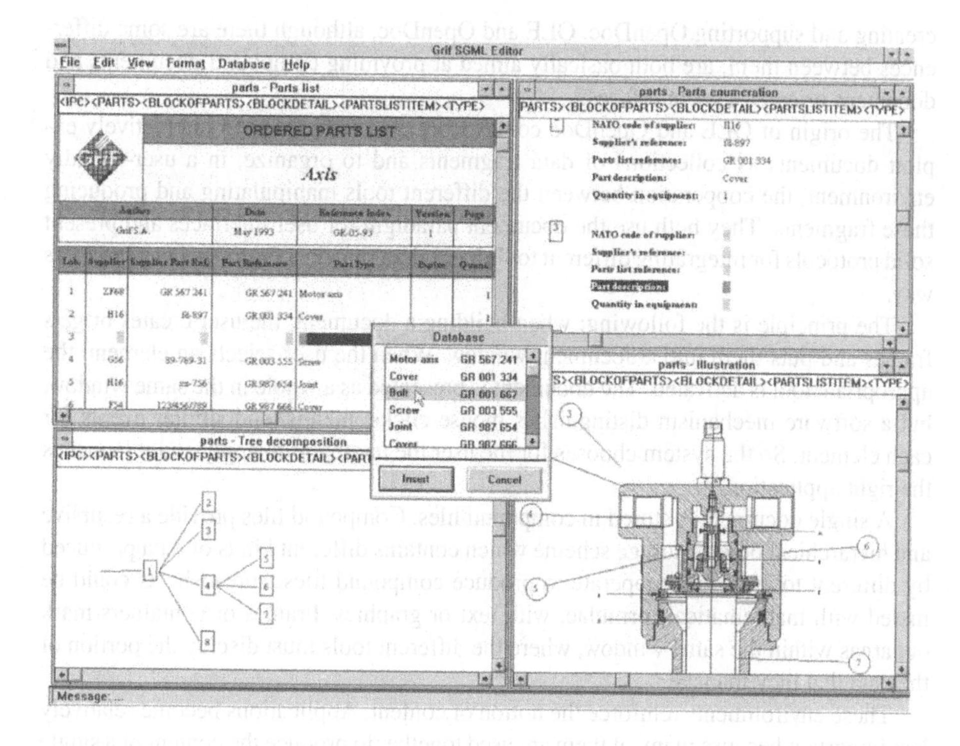

Fig. 3. An illustrated parts list catalog

In data sheets, which provide technical information about electronic components, characteristics are parametric data measurements such as conditions, values, signal names, test, circuit, and procedures. Characteristics data fragments are used in many different contexts and exploited by a wide range of tools, including databases and CAD/CAM applications as well as by document editing and viewing tools.

In the above examples, the integration of multiple tools is done in a seamless way for the end user: data is generated in the document and sent from tool to tool. The behavior of the document fragments is very specific to the data being manipulated and to the tools that manipulate them.

6 Related work

The concept of active structured documents has now proven its advantages and it is offered in more and more applications. It is also the underlying basis of some industry standards, such as OLE (Microsoft 1994) and OpenDoc (OpenDoc Design Team 1994). In this section, we briefly present these two standards and we introduce some other works that have been done in the field.

6.1 Compound documents

The user interface philosophy presented above has been endorsed by major companies: Microsoft is increasingly committed to OLE and Apple, IBM, Novell, and others are creating and supporting OpenDoc. OLE and OpenDoc, although there are some differences between them, are both basically aimed at providing compound documents and document-oriented user interfaces.

The origin of OLE and OpenDoc comes from a growing need to interactively exploit documents as collections of data fragments and to organize, in a user-friendly environment, the cooperation between the different tools manipulating and producing these fragments. They both use the document paradigm for user interfaces and present solid protocols for integrating different tools by exchanging data fragments in a seamless way.

The principle is the following: when building a document, the user creates or gets frames and puts them into a document window. When the user selects an element, the appropriate tool is activated. The document is presented as a whole in the same window, but a software mechanism distinguishes mouse clicks, menus, and similar events for each element. So the system chooses for the user the treatment to apply and it launches the right application.

A single document is stored in compound files. Compound files provide a recursive and hierarchical object-storage scheme which contains different kinds of data produced by different tools. Tools cooperate to produce compound files: spreadsheets could be mixed with mathematical formulae, with text or graphics. Frames or containers mark out areas within the same window, where the different tools must display the portion of the data that they manage.

These environments reinforce the notion of content. Applications become relatively less important because many of them are used together to produce the content of a single document.

OpenDoc, OLE, and SGML

OpenDoc and OLE give a vision of a document as the assembly of different data types coming from different environments. This vision is a very flexible one because it allows the user to aggregate with very little effort very sophisticated data coming from extremely specialized tools.

A sophisticated mechanism (properties and values) is used to allow multiple storage formats for each part. Tools can save each part in multiple formats (e.g., RTF, ASCII, and PostScript), in the hope that the person receiving the document possesses a tool that understands at least one of these formats, or perhaps is able to convert from one of these formats.

Another approach could be to use a standard exchange format for all parts of compound documents. These environments fit in perfectly with a tool-independent storage format such as SGML, and the coexistence of SGML and non-SGML data should be encouraged.

Exchanging structured information between tools

It would be unreasonable to imagine a situation in which all data in all the universe was instantly transformed into SGML. It would be more reasonable to encourage people to migrate only their most critical data to SGML, and to insure them that this data could be integrated seamlessly with noncritical data stored in proprietary formats.

In SGML, two types of markup can be distinguished:

- general-purpose markup identifies widely used document portions such as sections, subsections, paragraphs, lists, etc.;
- semantic markup identifies more precisely corporate- or industry-specific information such as motors, product parts, transistors, or other objects which require a very precise description (mathematical formulae, for instance).

In this case, SGML can benefit to environments such as OLE and OpenDoc, because data are more clearly identified by the semantic markup. As the system knows more about data, it can permit, when building OLE or OpenDoc applications, to identify more easily what tool to launch. This is why semantic markup constitutes the ideal fragment of information that can be produced, manipulated, and exchanged in OLE and OpenDoc.

6.2 Active documents

OpenDoc and OLE are clearly intended for business applications; but the academic community is also active in the field of active structured documents.

Some parallel work has been carried out at the University of Queensland, which has investigated the concept of active structured documents in a similar manner (Welsh 1994) as Grif. In the field of computer algebra system, CAS/PI (Kajler 1992, 1993) takes a software-engineering approach with a flexible and extensible system that also uses the concept of active documents.

Active documents are also considered in the ACELA project, which aims to build an environment for creating interactive books, specifically mathematical textbooks.

7 Conclusion

In this paper we have shown that active structured documents can be used as a paradigm for user interfaces. Active structured documents allow applications to be developed at a higher level of abstraction than with traditional user interface toolkits. They also allow very complex layouts to be handled interactively at a very low cost for the application. Several experiments have been made with various types of applications and various types of editors, but with a structured model of documents and generic definitions, active documents are especially well suited for manipulating mathematical formalisms.

The main advantages of active structured documents in the context of computer algebra environments can be summarized as follows.

The toolkit takes care of formatting and incremental redisplay of complex layouts. Using API functions, the application has only to create and update much simpler abstract trees.

Very rich graphical views can be generated, mixing multi-font text, graphics, tables, mathematical symbols, etc.

The graphical appearance of equations and other structured objects manipulated by mathematicians is expressed at a high level, as well as the interface between the editor and an application. Declarative languages allow applications to specify how each element type or attribute must be displayed and what kind of events concerning which element types or attributes they are interested in.

Acknowledgements

The authors thank the anonymous referees who have greatly helped to improve the original manuscript. The LaTeX parser presented in Sect. 5.3 has been developed at the Euromath Centre in Copenhagen in 1992 by Niels Jõrgen Kokhølm. Bjørn von Sydow is the main architect of the Euromath DTD and Lars Pedersen implemented the connection to MuPAD.

References

Adiba, M., Collet, C., Dechamboux, P., Defude, B. (1992): Integrated tools for object-oriented persistent application development. In: Tjoa, A M., Ramos, I. (eds.): Database and expert systems applications. Springer, Wien New York, pp. 439–445.

André, J., Furuta, R., Quint, V. (1989): Structured documents. Cambridge University Press, Cambridge.

André, J., Vatton, I. (1994): Dynamic optical scaling and variable sized characters. Electron. Publish. 7: 231–250.

Apple Computer, Inc. (1991): Inside Macintosh. Addison-Wesley, New York.

Arnon, D., Beach, R., McIsaac, K., Waldspruger, C. (1988): CaminoReal: an interactive mathematical notebook. In: van Vliet, J. C. (ed.): Document manipulation and typography: EP '88. Cambridge University Press, Cambridge, pp. 1–18.

Bier, E., Goodisman, A. (1990): Documents as user interfaces. In: Furuta, R. (ed.): EP '90: proceedings of the International Conference on Electronic Publishing, Document Manipulation and Typography. Cambridge University Press, Cambridge, pp. 249–262.

English, P. M. (1990): An extensible, object-oriented system for active documents. In: Furuta, R. (ed.): EP '90: proceedings of the International Conference on Electronic Publishing, Document Manipulation and Typography. Cambridge University Press, Cambridge, pp. 263–276.

Fuchssteiner, B., Gottheil, K., Kemper, A., Kludge, O., Morisse, K., Naundorf, H., Oevel, G., Schulze, T., Wiwianka, W. (1993): MuPAD Benutzerhandbuch. Birkhäuser, Basel.

Furuta, R., Quint, V., André, J. (1988): Interactively editing structured documents. Electron. Publish. 1: 19–44.

Harbo, K., Pedersen, L., Feddersen, H., Schaumburg, T. (1993): Interfacing Euromath to computer algebra systems: architecture and design. Euromath Report.

IBM (1991): Systems application architecture, common user access, advanced interface design reference. SC34-4290-00, First edn.

I.S.O. (1986): Information processing – text and office systems: Standard Generalized Markup Language (SGML). ISO 8879.

I.S.O. (1989): Information processing – text and office systems: Office Document Architecture (ODA). ISO 8613.

Kajler, N. (1992): CAS/PI: a portable and extensible interface for computer algebra systems. In: Wang, P. S. (ed.): Proceedings of the ACM International Symposium on Symbolic and Algebraic Computation (ISSAC '92), Berkeley, California. Association for Computing Machinery, New York, pp. 376–386.

Kajler, N. (1993): User interface for symbolic computations: a case study. In: Proceedings of the ACM Symposium on User Interface Software and Technology (UIST '93), Atlanta, Georgia. Association for Computing Machinery, New York, pp. 1–10.

Knuth, D. E. (1984): The TeXbook. Addison-Wesley, Reading, MA.

Maraninchi, F. (1991): The argos language: graphical representation of automata and description of reactive systems. In: Proceedings of the IEEE Workshop on Visual Languages, Kobe, Japan. IEEE Computer Society Press, Los Alamitos, CA.

Microsoft (1992): Win32 application programming interface. Microsoft Press, Redmond, WA.
Microsoft (1994): OLE 2.0 programmer's reference, working with Windows objects. Microsoft Press, Redmond, WA.
Open Software Foundation (1992): OSF/Motif programmer's reference, revision 1.2. Prentice-Hall, Englewood Cliffs, NJ.
OpenDoc Design Team (1994): OpenDoc, the required reading packet.
Paoli, J. (1995): Cooperative work on the network: edit the WWW! Comput. Networks ISDN Syst. 27: 841–847.
Quint, V., Vatton, I. (1986): Grif: an interactive system for structured document manipulation: In: van Vliet, J. C. (ed.): Text processing and document manipulation, proceedings of the international conference. Cambridge University Press, Cambridge, pp. 200–213.
Quint, V., Vatton, I. (1987): An abstract model for interactive pictures. In: Bullinger, H.-J., Shackel, B. (eds.): Human-computer interaction, Interact '87. North-Holland, Amsterdam, pp. 643–647.
Quint, V., Vatton, I. (1994): Making structured documents active. Electron. Publish 7: 55–74.
Quint, V., Roisin, C., Vatton, I. (1995): A structured authoring environment for the World-Wide Web. Comput. Networks ISDN Syst. 27: 831–840.
von Sydow, B. (1992): The design of the Euromath system. Euromath Bull. 1/1: 39–48.
von Sydow, B. (1994): Editing mathematics in the Euromath system. Euromath Bull. 1/2: 17–23.
Terry, D. B., Baker, D. G. (1990): Active Tioga documents: an exploration of two paradigms. Electron. Publish. 3: 105–122.
Timoney, R. (1994): The construction of an interactive LaTeX translator for mathematical formulae. Euromath Bull. 1/2: 103–110.
Welsh, J., Han, J. (1994): Software documents: concepts and tools. Software Concepts Tools 15: 12–25.

Direct manipulation in a mathematics user interface

Ron Avitzur

1 Introduction

The user interface problems of existing mathematics systems are well known and are discussed in detail elsewhere (see, e.g., Kajler and Soiffer 1998).

Mathematics systems share the complexities of all large modern software: hundreds or thousands of individual commands, countless options, and hundreds of pages of documentation. Few users of word processors learn a substantial fraction of their features. Mathematics systems compound this complexity by using non-standard linearized syntaxes for typing mathematical and command lines with custom programming languages.

The Graphing Calculator is an experiment to see how much functionality can be available in a system without command lines and minimizing the reliance on buttons, menus, and dialogs. Since Apple includes it with every Power Macintosh, ease of use was the overriding design constraint. We cut features rather than present them awkwardly. We strove for a design visually and functionally minimalist to avoid intimidating users.

This paper describes the design decisions made in the Graphing Calculator on Power Macintosh. Section 2 covers previous work both in other systems and in prior packages from which the Calculator is derived. Section 3 discusses our design goals and guidelines. Sections 4 and 5 describe the static and interactive interface elements, respectively. Section 6 discusses usability in various forms. Section 7 is on the use of interval arithmetic in plotting. Section 8 briefly touches on implementation details. Section 9 outlines some directions for future work.

2 Previous work

2.1 Other systems

Many systems today have two-dimensional structured expression display, entry, and editing. Kajler and Soiffer (1998) provide an excellent survey of user interface work in computer algebra systems. All the user interface elements in the Calculator exist in some form in other systems. Milo (Avitzur 1988) and Theorist (Bonadio 1989) both have types of direct manipulation on equations and graphs. WinMath is an interactive graphing system with both more functionality than the Calculator and a correspondingly more complex user interface (Gourdol and Lafourcade 1987). Maple, Mathematica, and Axiom produce beautiful graphics, though in a less interactive manner. We examined both the functionality and user interface of many systems in the design process. The classroom use of TI and HP graphing calculators were particularly instructive (Demana and Waits 1988).

2.2 Product history

The Calculator embeds the typesetting, editing, and algebra system of Milo and FrameMath. Milo is a Macintosh application which integrates text, equations, and pictures into documents with a toy symbolic mathematics system tailored to the problems in undergraduate physics (Avitzur 1988). Milo's user interface has many weaknesses. Keyboard shortcuts use up to four keys pressed at once. The commands are difficult to find in long hierarchical menus. Nonetheless, it was one of the first systems in which expressions are both displayed and edited in the standard two-dimensional notation. Once the necessary keyboard shortcuts are mastered, it is faster to edit expressions than with a linearized syntax. Milo has primitive 2-D graphing. One can use the mouse to zoom in on interesting regions of a graph.

The word processing of Milo is extremely primitive, so the code was then embedded in FrameMaker as the FrameMath mathematical typesetting engine (Avitzur 1989). This version improved the interface with a graphical palette of buttons and improved mathematical typography. FrameMaker is a long-document desktop publishing product. It is the only page layout program capable of symbolic differentiation.

The Graphing Calculator began with this formula editor and symbolic-computation system as a code base. Thus many of the design decisions were not a question of what to implement, but rather which features to remove to simplify the interface.

The project was an independent commercial effort of its authors, Ron Avitzur and Greg Robbins, in collaboration with Apple Computer. We began July, 1993 and took six months. The basic design and functionality were completed in one month so that most of our time could be dedicated to quality assurance, usability testing, and step-wise refinement of the user interface. Because the Calculator is included on the hard disk of every system sold, our schedule was tied to the hardware manufacturing ramp and was not flexible; if we were late, we would not ship. Because the Calculator goes to every Power Macintosh user, we maintained extremely high standards of both reliability and usability.

3 Design overview

3.1 Goals

Ease of use for inexperienced users is a strict filter of design choices, but not a productive criterion for generating design options. Our primary design goal was to provide a tool useful in high-school education. We chose our features on their pedagogical value rather than computational power. To this end, we examined high-school algebra books and attempted to address a wide class of their problems.

Command line interfaces surely, and most computation systems generally, put the user in an interaction mode of several steps: query, (pause), examine the response, consider how to modify the query, repeat. If a computation takes a long time, the user considers it expensive and will try to make as few queries as possible. A pause may be inserted either by the calculation time, or by a complex interface. This breaks the user's flow of thought and causes her to spend more time thinking how to use the software, rather than concentrating on the mathematics.

We strove to make the Calculator so responsive that users perceive actions as inexpensive, so that the mental model of interaction is query, query, query, ... going through variations quickly, even frivolously, and focusing so totally on the mathematics as to pay no attention to the tool. Our usability design criterion was to make interaction so fluid and natural that users could play with the mathematics as if it were a game.

As usual in commercial projects, time constraints limited our design. Because our deadline was inflexible, we took no risks and had little time to experiment. We used only well-understood methods which did not require lengthy development or debugging.

3.2 Usability guidelines

Our guidelines were simple. Ideally, the user should not need to learn anything about the program, or make any choices directing the software. We desired to focus the user's full attention on the mathematics rather than the computer. Ideally, we wished to have no buttons, menus, dialog boxes, or command lines. With this in mind, whenever we added a menu item, dialog, or control, it had to greatly improve the pedagogical value, or clearly improve the usability as demonstrated by user testing. As a result, the user is given very few choices. The method of rendering surfaces and the fonts and layout used in graphs cannot be changed. We felt that the user interface complexity added to put these under user control was not justified because these features did not add pedagogical value to the product.

Usability testing demonstrates the value of redundant encoding. User actions ought be doable in several ways, perhaps as a menu item with a keyboard shortcut as well as a button on a palette and directly with the mouse. This contradicts our desire for a minimal interface by adding multiple controls for the same action. Yet, in the user's mind, the interface is simple if the number of actions are small. Once one way to do something is learned, others are ignored if the graphic design is visually uncluttered. Hence, we kept low the number of program "verbs" the user needs to learn.

In many places, we use direct manipulation to provide an interface to functionality without requiring the user to learn a new "verb." The user can change the domain or perspective view of a graph, animate a family of curves, solve an equation in one variable, and algebraically manipulate an equation the same way she changes the size of a window pane by grabbing and dragging an object with the mouse. The following sections explain how these diverse actions are unified under a common interface paradigm.

4 Overview of static interface elements

The first thing a new user sees is unintimidating and visually simple (see Fig. 1). We wished to maintain the simplicity of the original Calculator desk accessory (Fig. 2) for arithmetic operations. Arithmetic results appear to the right of the expression (Fig. 3).

Figure 4 shows the interface elements in the main window. The window is divided into two panes: the math pane for editing expressions and the graph pane for displaying and manipulating graphs. The pane divider can be dragged to any position. The zoom buttons appear only when the graph is visible and control the scale of the axes in discrete

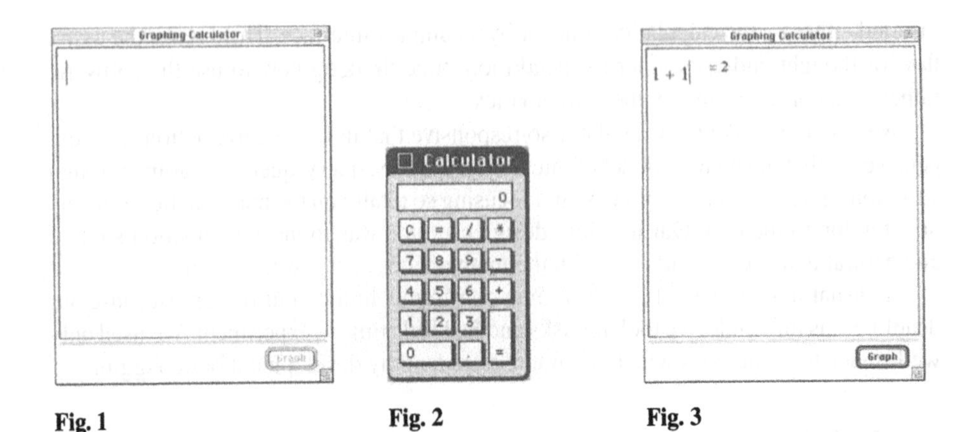

Fig. 1. A visually simple first impression
Fig. 2. The original 1984 Calculator desktop accessory
Fig. 3. Arithmetic computations

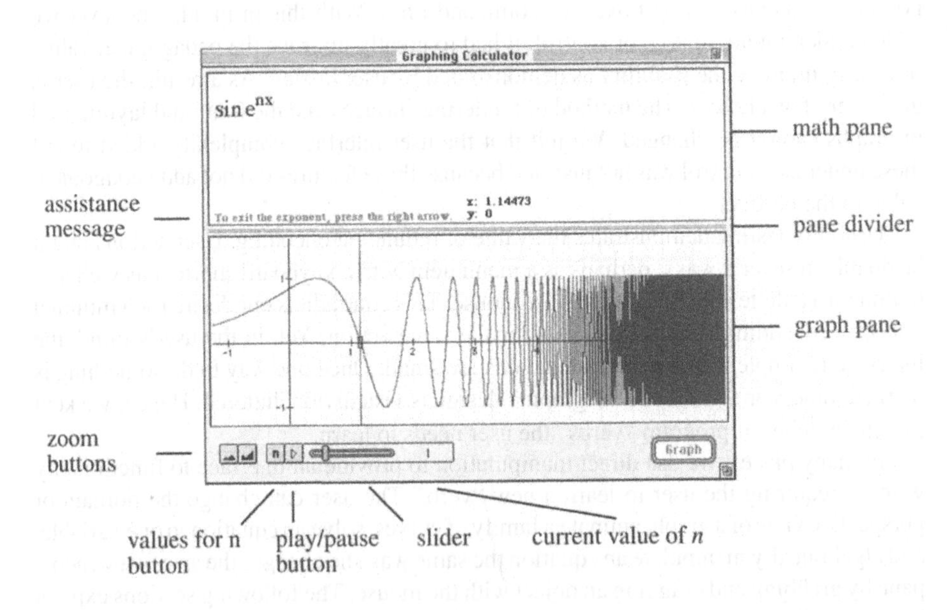

Fig. 4. Interface elements

steps. The slider appears only when n occurs in the expression and controls the numeric value for n. The n button brings up the dialog in Fig. 7 to specify the range of values for n. The play button causes the slider to automatically vary in order to animate the graph.

Figure 5 shows all the menus. The functionality is also available from the palettes in Fig. 6. The File menu is discussed in Sect. 5.5. The Edit menu is standard across applications except for the Copy Graph command discussed in Sect. 6.5. The Equation menu, which is redundant with the large keypad, contains commands for typing math-

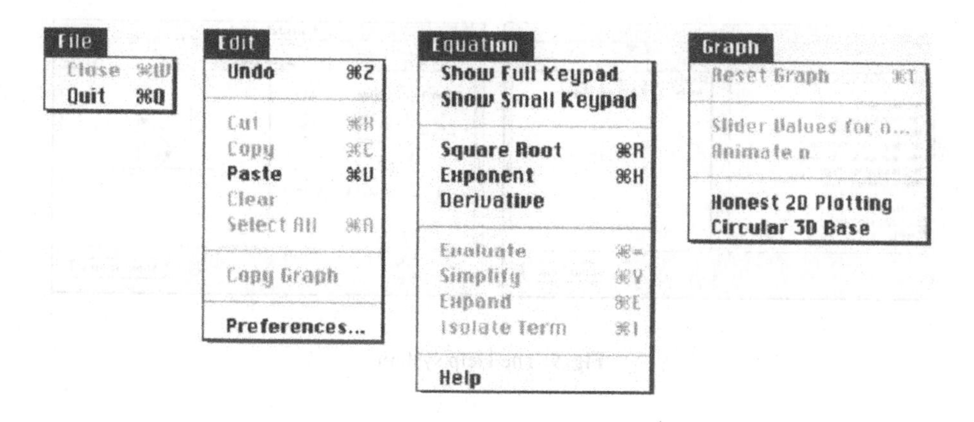

Fig. 5. The menus

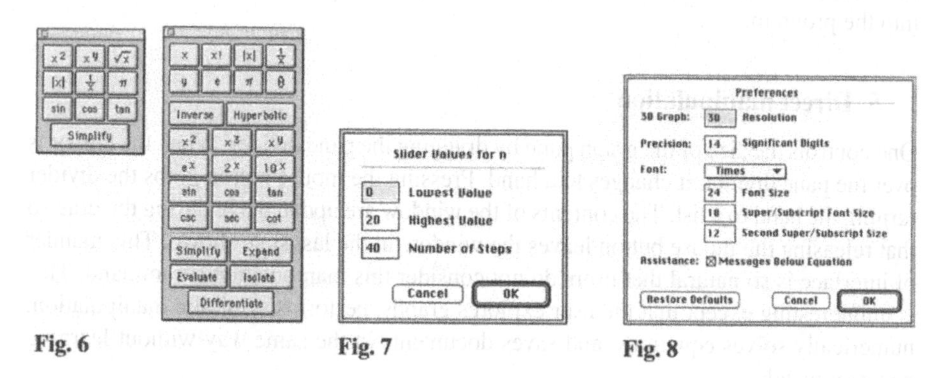

Fig. 6 **Fig. 7** **Fig. 8**

Fig. 6. The palettes
Fig. 7. A dialog controlling the parameter *n*
Fig. 8. The Preferences dialog

ematics and symbolic manipulation. We included the small keypad because we felt the full version too complex for young users (Fig. 6). The Help system (Fig. 9) is a typical browser. The Graph menu provides a redundant encoding of the buttons attached to the slider, as discussed in Sect. 5.4. "Reset Graph" changes the scale and moves the origin to default values. "Honest 2D Plotting" is a global mode discussed in Sect. 7. "Circular 3D Base" is a global mode toggling the domain over which 3D surfaces are plotted between a circle and a square so that radial symmetry may be more visually apparent. The Preferences dialog (Fig. 8) is an embarrassment given our design goals, however, we were unable to agree on universally acceptable parameters. We hope typical users will be satisfied with default values and will not use the dialog. These are all the menus, dialogs, and buttons. Many are redundant; users can do most things without using any of these interface elements.

Because this system is extremely interactive and responsive, it is difficult to convey

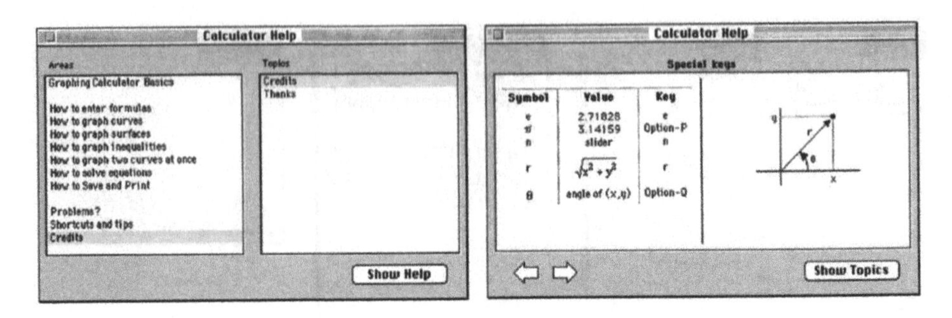

Fig. 9. The Help system

on paper. The effectiveness of the interface comes from one's ability to interact with it quickly. One can get a better understanding of the interface by finding a Power Macintosh (the software is on every machine) and watching the self-running demonstration built into the program.

5 Direct manipulation

One controls the size of the graph pane by dragging the pane divider. When the cursor is over the pane divider, it changes to a hand. Pressing the mouse button grabs the divider turning the hand to a fist. The contents of the window are updated live during the drag so that releasing the mouse button leaves the window in the last state shown. This manner of interface is so natural that users do not consider this manipulation a command. This is uninteresting except that the user explores graphs, performs symbolic manipulation, numerically solves equations, and saves documents in the same way without learning new commands.

5.1 Direct manipulation in graphing

After entering an equation, pressing the graph button immediately produces a picture. There is no dialog box querying the domain, range, variables, or other options. The graph can update as fast as the user edits the equation. To simplify the user interface, the user has virtually no control over the plotting. We compute one point per screen pixel to avoid the need for options related to adaptive plotting. Rather than require numeric entry of the domain beforehand, a default view is chosen. The user can then pan and zoom using the mouse. Figure 10 shows a sequence of panning and zooming without showing the dozens of intermediate frames. The zoom buttons change the scale, showing several intermediate scales to animate the zoom so the user maintains a sense of place. When the cursor is over the axes, it becomes a hand, indicating that the axes can be grabbed. Clicking the mouse button changes the hand into a fist grasping the axes. While dragging the axes, the function is computed along the edges and the axes redrawn without flicker to give the impression that one is merely moving the window pane over a larger piece of graph paper. The placement and icons of the zoom buttons are common in Macintosh applications to zoom in or out on a document. Since the tick marks and labels on the axes are animated during the zoom even if there is no graph, the function of the buttons

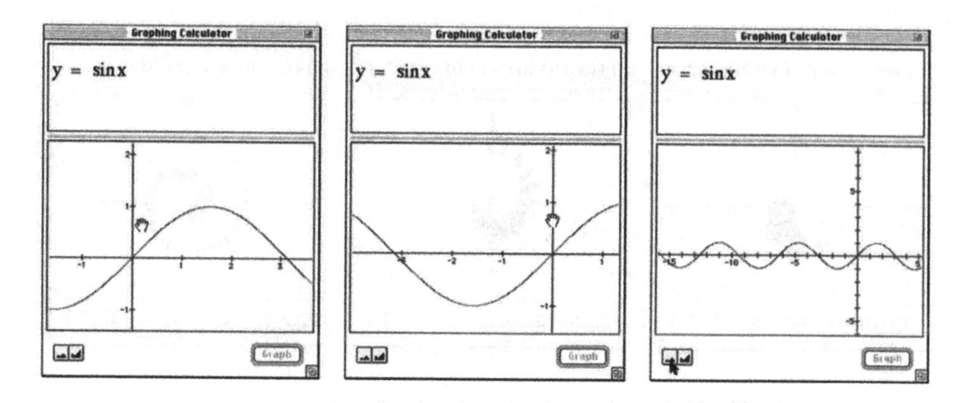

Fig. 10. Panning & Zooming to control the domain

are obvious to a new user. (Clicking anywhere on a 3-D graph will spin the surface. In order to change the domain, one must hold down the option key to pan. We do not expect users to discover this without reading the documentation, but neither do we feel that it is important.)

The user does not specify the graph variables. Rather, the form of the equations which can be graphed is restricted to the forms we found typical in high-school algebra texts. The Graphing Calculator will graph equations of the form:

$y = f(x)$,
$z = f(x, y)$,
$f(x) = g(x)$,
$f(x, y) < g(x, y)$.

Functions of the form $f(x)$ are automatically graphed in two dimensions, while functions of the form $f(x, y)$ are automatically graphed in three dimensions. In addition, (r, θ) can be used in place of (x, y) in 3-D graphs and inequalities. Using these conventions, it is necessary neither to specify a coordinate system nor whether a 2-D or a 3-D graph is desired. Students using the Calculator for the first time were able to open their algebra book to a random page, type in a formula and see the appropriate graph immediately with no instructions.

5.2 Direct manipulation in viewing a family of curves or surfaces

Since students can develop an understanding of functions by exploring, we made the Calculator as responsive as possible. To further simplify common explorations such as $\sin x$, $\sin 2x$, $\sin 3x$, the letter n in the Calculator is a special constant. If it occurs in an expression, a slider appears at the bottom of the window and n takes the value specified by the slider. As the user moves the slider with the mouse, the value of n changes and the graph is immediately replotted. By changing n automatically, an animation can be conceptualized as a single object representing the family of curves. Figure 11 shows a few frames of an animation of inequalities in polar coordinates.

Following our design principles, there is only one slider and although the letter n

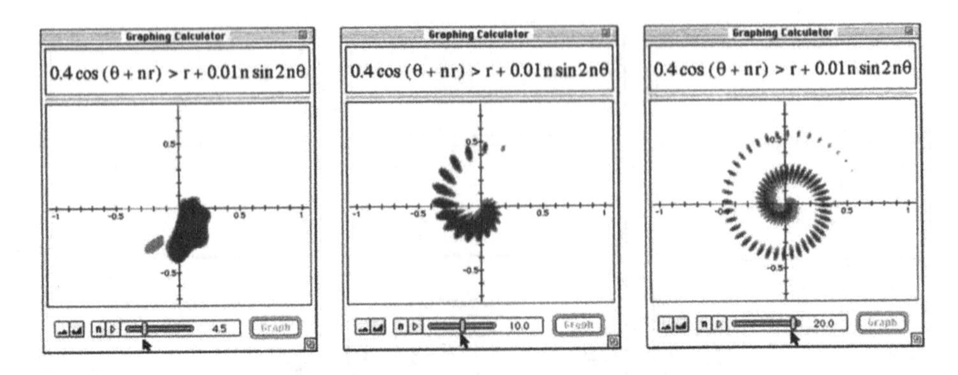

Fig. 11. Varying n animates a family of implicit graphs

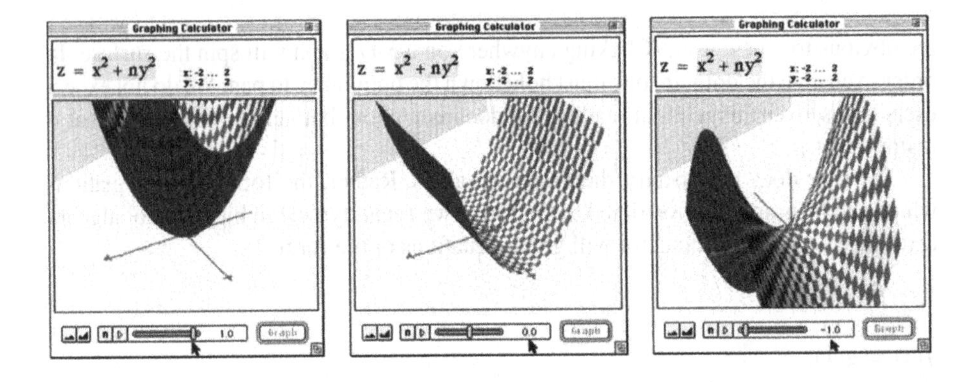

Fig. 12. Varying n animates a family of surfaces

was chosen arbitrarily, the user has no choice in the matter. A dialog box specifies the range of values for n.

Figure 12 shows slices of an animation illustrating that a paraboloid, a cylinder with parabolic cross section, and a saddle are instances of a family of surfaces parametrized by n. This demonstrates visually the difference between positive and negative Gaussian curvature.

5.3 Direct manipulation in numeric equation solving

The curve is itself an active region to query the numeric value of points on the graph. Clicking on the curve brings up a vertical line marking the graph and feedback showing the coordinates. That line can then be dragged back and forth. (In a 3-D graph, one cannot inspect the values. Clicking anywhere will spin the graph under control of the mouse.)

Figure 13 a shows that $\sin x = x/10$ has 7 solutions, all in $-10 < x < 10$. Clicking on the curve marks that point and displays its coordinates. Dragging the vertical line inspects the value of the curve. At ordinary points, x corresponds to that screen pixel

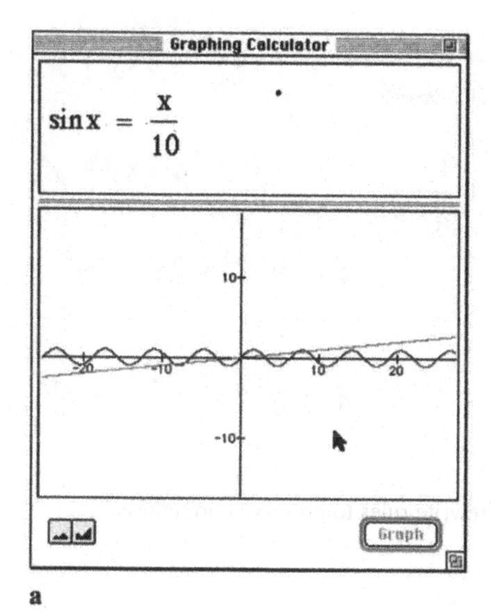

a

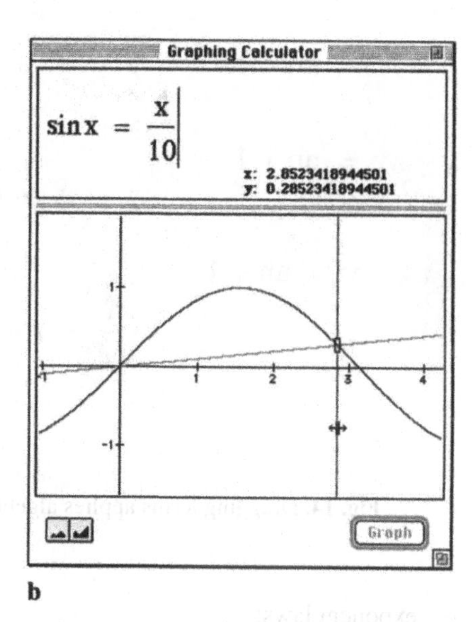

b

Fig. 13. **a** Seven solutions to $\sin x = x/10$, **b** clicking near one automatically invokes a root-finder

so as the mouse is moved, x steps in increments of one pixel. At interesting points (zeros, solutions to the equation, local minima, and local maxima) along the curve, the vertical line visibly and audibly snaps to the interesting point (see Fig. 13 b). As the user drags the mouse, the calculator evaluates many nearby points to look for anything interesting. If something interesting is nearby, it numerically finds its coordinates to machine precision. Since numeric methods for finding extrema are good to only half the machine's precision, it finds a zero of the symbolic derivative to find the extrema. These calculations are behind the user's back. Merely clicking on the curve and dragging a marker will display useful values without having to explicitly invoke a numeric method.

5.4 Direct manipulation in symbolic algebra

Using the mouse to move objects or pan a view of a document is quite old and was popularized to a large audience when Apple bundled MacDraw and MacPaint with the original Macintosh in 1984. The feature in the Calculator which delights the most people reduces the rote manipulation one learns in high-school algebra to the same paradigm of dragging boxes used in MacDraw. The user's mental model is the same as in a drawing program. One selects an object and moves it. In this case, one selects a subexpression and with the hand cursor which appears and drags it back and forth. The program performs the necessary symbolic manipulation to preserve the correctness of the expression. The columns in Fig. 14 demonstrate this. Of course, the steps are reversible. Some of the rules it uses are:

- commutativity of addition and multiplication to swap adjacent terms;
- distributive law of multiplication over addition both to factor and multiply out terms;

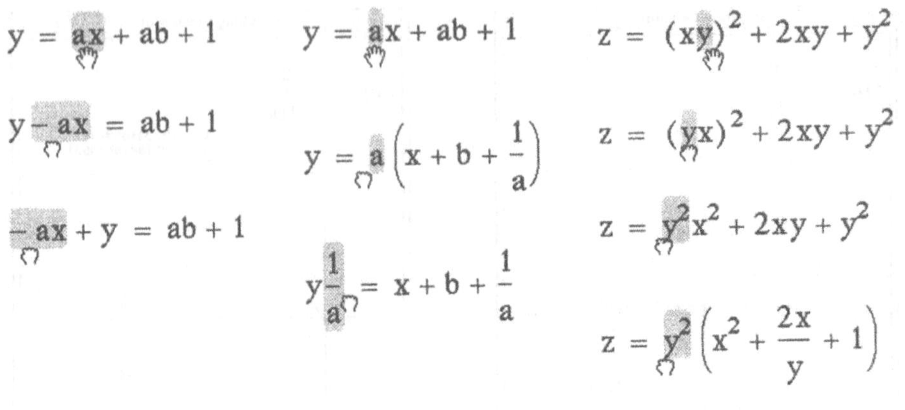

$$y = ax + ab + 1 \qquad y - ax = ab + 1 \qquad -ax + y = ab + 1$$

$$y = ax + ab + 1 \qquad y = a\left(x + b + \frac{1}{a}\right) \qquad y\frac{1}{a} = x + b + \frac{1}{a}$$

$$z = (xy)^2 + 2xy + y^2 \qquad z = (yx)^2 + 2xy + y^2 \qquad z = y^2x^2 + 2xy + y^2 \qquad z = y^2\left(x^2 + \frac{2x}{y} + 1\right)$$

Fig. 14. Dragging terms applies algebraic rewrite rules to preserve correctness

- exponent laws;
- subtract or divide a term from both sides to move it across an equal sign.

5.5 Direct manipulation in lieu of document commands

Following the user interface model of Macintosh Desk Accessories, the File menu has only two items: Close and Quit. We do not implement New, Open, Save, Save As, Page Setup, or Print, which are standard in applications with documents. We simplified the interface by eliminating documents. To print, one must copy pictures to a word processor. One can save state using direct manipulation. If the Macintosh Drag and Drop system extension is installed, one can drag graphs or equations onto other applications which support the drag manager and accept "PICT" objects. Dragging onto the Macintosh Finder creates a clipping file with the equation or graph. Dragging that clipping back onto the Calculator pastes the contents into the Calculator.

6 Usability

6.1 Usability testing

During development, we refined the interface with informal testing bringing new people to our office every day and silently watching them use the latest version. Near the end of development we performed formal user studies with a professional in usability testing. Teachers and students in groups of two and three for periods of 90 minutes had both specific tasks and time for exploration of the system. They were given no instructions in the use of the system. Their interaction was both videotaped, and viewed live from behind a one-way mirror. We watched users stumble over assumptions we took for granted.

Features of the windowing environment such as the system location for application Help, or even how to resize windows posed problems. Subtle aspects of our system were overlooked as users focussed on the simplest tasks. Users employed a common approach to problem solving. They would read each menu item, examine each control in the window, and stare at each button on the palette very slowly. However, they were hesitant to try things and would sometimes look at the appropriate control and move on to the next. Users did not spontaneously use the on-line help. When prompted to do so, we watched in horror as users spent 15 minutes trying to find the Help. Macintosh guidelines specify that application Help belongs in a system menu in the right corner of the menu bar. Users do not know this. They stared at the Equation menu for a long time, so we put the Help there. Everyone browsing Help read "Shortcuts and Tips" first, so we put the most important information there since it may be the only thing some users read.

In formal user studies, users quickly discovered and understood the zoom buttons. They did not, however, readily discover that the axes are a control for panning. We did not give the axes the visual appearance of a control for aesthetic reasons. Changing the cursor is too subtle a cue for typical users. Users in the study did stumble on how to pan after several minutes of experimentation. This control is something which seen once is not forgotten. We added a self-running demonstration mode to the program which shows the cursor moving slowly, with explanations, and demonstrates everything which can be grabbed. In a 90-second demonstration, users quickly learn the programs controls. We felt this so important that an entire menu is devoted to self-running demos.

It is often easy, and certainly gratifying, to program powerful algorithms and functionality. In contrast, since everyone has a different notion of what is natural, it can be expensive, slow, and frustrating to make software usable for a large audience. We chose not to implement many features because for each piece of functionality we took a long time to refine its interface by testing with many people. Such an approach is difficult in a marketplace where products compete on feature checklists resulting in products with thousands of features which few customers use.

Programmers are rarely trained in usability issues. Since people tend to design systems for themselves, programmers, being necessarily expert users and often having contact only with other expert users, are the least qualified to design user interfaces for typical users.

During the design process, we often would agree on obvious ways to simplify the interface. Other times our intuition would differ and we could not from general principles decide on the right solution. Ultimately, real users show us what is simple or difficult.

6.2 Assistance

We classified many of the users difficulties as program bugs and fixed them. Other times, although the program's behavior is unexpected, changing it would lead to unexpected behavior in other circumstances. For example, typing "x^2+1" produces x^{2+1}. The insertion point remains in the exponent. If we automatically drop the insertion point out of the exponent, typing x^{10} would be difficult. If we decide based on precedence whether the next character remains in the exponent, as in the "Fortranish" mode of Theorist, then the insertion point could not give reliable feedback, and the behavior would be more complex for a user without expectations based on other systems. We addressed this issue by displaying assistance messages in circumstances users may find confusing. When the insertion point is after an exponent, the message "To exit the exponent, press the right arrow key." appears. This message is in light gray at the bottom of the pane, so users typically do not notice it. Yet, when something unexpected occurs, they will stop and their focus of attention will widen. If a solution is already on the screen, they continue their task with minimal delay. User testing showed us circumstances in which the program's behavior was not obvious. The software looks for these cases and displays an appropriate message. Users found this both natural and helpful.

6.3 Graphic design

We cannot overstate the importance of professional graphics design to make a good interface. The precise layout and visual appearance of the window, palettes, controls, and feedback information is extremely significant. At first consideration, one may presume that the functionality and interactive quality is unaffected by superficial changes or that visual appearance is a window-dressing applied at the end of the design. In practice, visual noise, disorganization, or clutter is extremely damaging, particularly to inexperienced users. In formal user testing, we found it common for users attempting to solve a problem to slowly consider each object on the screen. The simpler the layout, the more quickly one can examine it. To quote Tufte (1990): "Confusion and clutter are failures of design, not attributes of information."

We use translucency for aesthetic effect in two places. Curves (which may be represented by solid patches where they oscillate quickly) are translucent, allowing the axes and labels to be seen through them. A selected point on a curve is marked with a translucent blue box showing the curve through it. This effect is more expensive to render than a traditional exclusive-or, but is visually more appealing.

6.4 Seeing is remembering

In formal usability testing we discovered that users did not recognize many controls as something to click on, such as the Play button, the *n* button, the pane divider, or the graph axes. These problems were addressed by adding those actions as menu items or demonstrating the controls in the self-running demo. After once seeing the pane divider or graph axes grabbed, no one forgot how to manipulate them. Although they were not recognized as controls, users did stumble across them by accident. By presenting them in an animated way automatically, the learning curve is reduced.

6.5 User interaction model

As discussed in our design goals, we want the user's mental model of interaction to be one of playing with the mathematics rather than a query-(pause)-response model of interacting with the program. This makes responsiveness and speed a top priority. Also to this end, we consider the user's sense of place extremely important. The graphing domain never changes automatically. It is always the same as the last graph (even across sessions), and only changes under direct user control of panning or zooming. With similar motivation, the graph is not computed unless the user explicitly chooses Graph. The machine is fast enough that it could graph the new expression in between each character the user types, thus eliminating the need for a Graph button, reducing the number of controls and the visual clutter. However, the resulting flicker would be distracting and could reduce the user's sense of control.

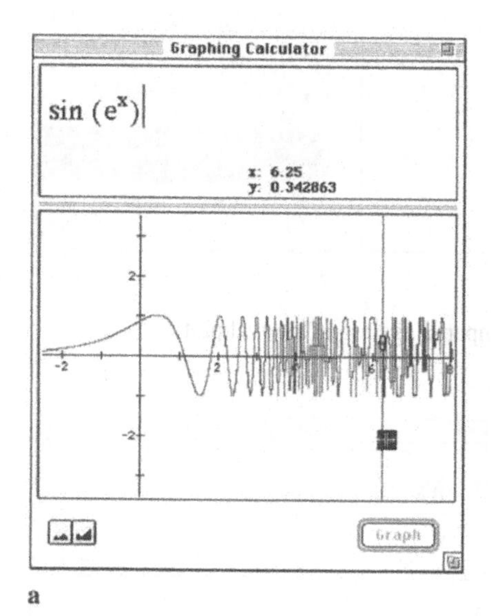

a

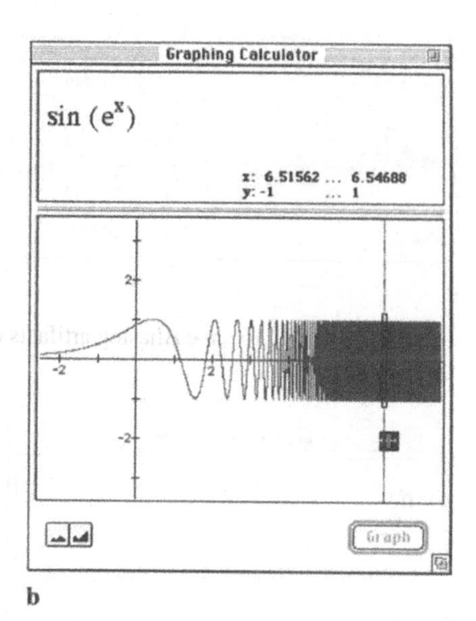

b

Fig. 15. a Noise on the right is due to undersampling. **b** Using interval arithmetic produces a cleaner picture

Another aspect of the interface which simplifies the user's mental model of interaction is that there is only one keyboard input focus. All typing goes to the equation. We considered eliminating the "Values of n ..." dialog and menu item by using two numeric edit fields in the main window. We decided against this both to reduce visual clutter in the window, and also because having multiple input fields requires the user to be aware of where typing is sent and how to redirect it. As a result of this decision, we needed a separate menu command, Copy Graph, because the Graph is never the input focus on which the Copy command acts.

7 Interval arithmetic

The Calculator optionally uses interval arithmetic to plot 2-D curves, following suggestions from Fateman (1992). This is a user interface issue for pedagogical reasons. Naive plotting algorithms can produce incorrect graphs which are quite misleading. We cannot expect our users to understand the plotting algorithm in order to make effective use of the software. Figure 15 a shows a smooth graph degenerating into sampling noise. Figure 15 b uses honest plotting to produce a better representation of the curve.

Extremely simple curves, of the form $\sin(kx)$ produce absolutely incorrect pictures. Figure 16 a shows the beat frequency between the frequency of the mathematical function and the frequency of the screen pixels on which the function is linearly sampled. The graph in Fig. 16 b is worse; at first glance it looks reasonable, yet it is entirely wrong. In signal processing, this effect is called aliasing. The honest plot is shown Fig. 16 c. The Calculator, in the examples without honest plotting, is sampling the function at one point

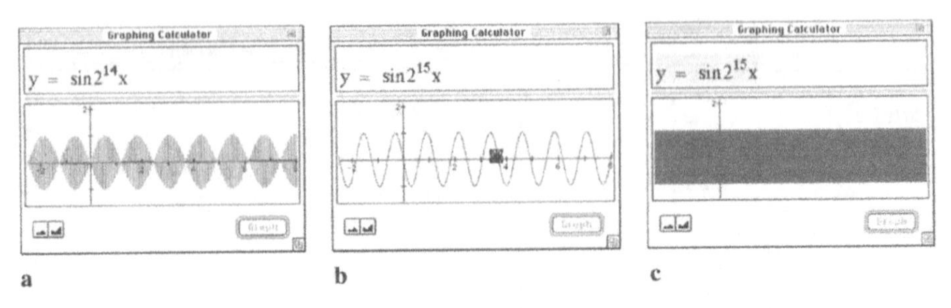

Fig. 16. a–c Aliasing artifacts compared against an interval plot

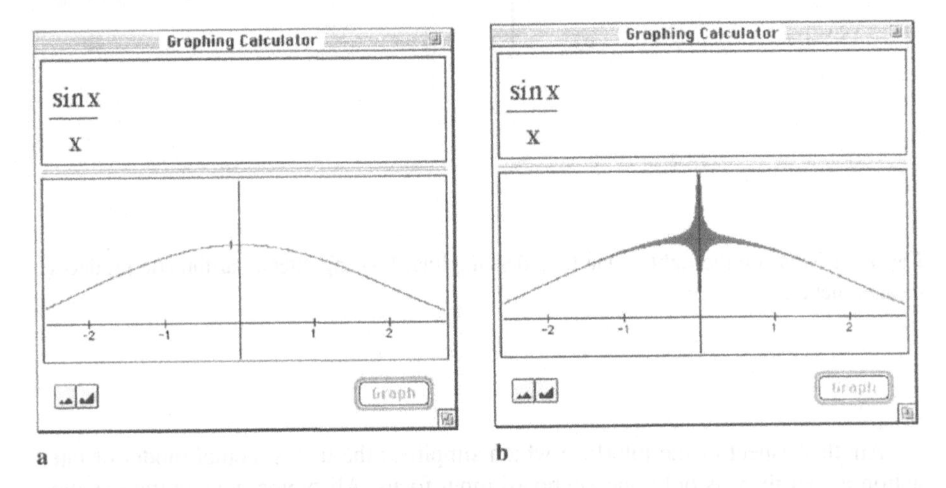

Fig. 17. a Linear sampling. **b** Pessimistic intervals are misleading

per pixel uniformly spaced. Some systems use adaptive plotting (Maple, Mathematica) while others may choose the sampling points randomly (Theorist) so that such simple examples don't fail dramatically. Any graphing algorithm based on sampling may be fooled by an adversary function which at all sampling locations returns 1, yet behaves arbitrarily elsewhere.

The Calculator deals only with compositions of elementary functions, so every function has a straightforward interval extension. Rather than sampling at discrete points, we evaluate the interval extension over an x-interval corresponding to the width of each pixel. The result is treated as a vertical error bar representing a bound on the curve. (It is an implementation bug that this bound is not rigorous because the rounding direction is not set prior to each floating-point operation.)

Unfortunately, the honest plotting mode also has failure modes which can mislead students. Consider Fig. 17 a and b which shows $(\sin x)/x$ graphed by linearly sampled points and by intervals. The honest plot bounds the correct graph, but is too pessimistic and implies that something interesting may happen near $x = 0$. We chose linear sampling as the default method since those failure modes are easier to explain to a high-school student.

Honest plotting is a global mode. If both methods produce the same picture, then the function is represented well, otherwise one must seek to understand the function further analytically. We did not like, aesthetically, the result of displaying both an honest and a naive plot superimposed. We are dissatisfied leaving this as a user option. Perhaps an adaptive method which plots the curve using intervals and then recursively subdivides intervals larger than one pixel could produce honest plots which are never too pessimistic. We did not have time to explore this.

We also use the interval arithmetic to compute inequalities as in Fig. 11 efficiently. Rather than evaluate the function explicitly at each pixel, we evaluate the interval extension over recursively subdivided rectangular intervals. If the expression is true or false over large areas, this is determined immediately. The inequality is rendered in three colors representing true, false, and maybe. The inequality may be both true and false over the extent of a pixel, or the interval arithmetic may be give pessimistic bounds, which we represent as an intermediate shade of color. This method of rendering implicit curves in two dimensions is discussed by Mudur and Koparkar (1984) and by Suffern and Fackerell (1991).

8 Implementation details

The Calculator is roughly 50,000 lines of primarily C code, except for the 3-D rendering package which is in C^{++}. The typing, display, and editing of 2-D mathematical expressions is derivative of Milo and FrameMath. This code is virtually identical to what is embedded in FrameMaker IV, except for changes suggested by usability testing. Soiffer (1991) describes in detail the issues involved in such 2-D mathematical structure editors, as well as this system in particular. The algebra system in the Calculator is loosely coupled to the structure editor, sharing data structures, memory management, and tree manipulation routines. The structure editor and the algebra system make up more than 50% of the Calculator. Since they were already well tested in other shipping products, we made as few changes as possible to them to avoid revalidating the code.

The direct manipulation of symbolic subexpressions is implemented as two commands: MoveRight and MoveLeft. These are rewrite rules based on the context of the subexpression hard-coded in C for efficiency. Today, efficiency would not be an issue, but the symbolic-manipulation library was written in 1986 for an 8 MHz 68000 Macintosh with extremely poor system memory management. As a result, the algebra system explicitly allocates and frees all memory, rather than using garbage collection or reference counting. This is faster, but extremely sensitive to pointer bugs.

The Calculator is a standard Macintosh application structured around an event loop. Calculations are done at idle time between user actions in 0.25-second chunks so that no matter how long a graph takes, the system is responsive to user events. For example, the inequalities of Fig. 11 may take a minute to compute. Every quarter second the screen will update to provide feedback and one can pan the graph while it is being drawn. Portions already computed remain and revealed regions fill in with the rest of the graph.

The code for graphing is not well structured. The authors intend to re-architect it with firewalls separating different types of graphs into modules and abstracting the interaction from the rendering. The most interesting algorithms we use in graphing involve interval arithmetic and are described in Sect. 7.

9 Directions for future work

The Calculator is a toy system and is best thought of as a user interface prototype. It has considerably fewer features than even the fifty-dollar hand-held graphing calculators. Given sufficient time and resources, we would like to match the features of other systems, while preserving the initial ease of use and allowing for a smooth learning curve. The notion of progressive disclosure is to hide complex functionality until the user is proficient with the tool.

Endless improvements of the strictly commercial kind are obvious and uninteresting. These include, to name a few items on our list, saving animations as QuickTime movies, calculations on complex numbers, supporting the forward delete key in mathematics editing, graphing more than two curves at once, and speed improvements of up to 100× which could be reached by compiling the typed expression to machine code rather than walking the tree to compute each point. The challenge is to extend the power of the system without making it more complex to use. One extension without changing any interface elements is to graph more forms, such as polar plots, 3-D surfaces in spherical coordinates, or inequalities in three dimensions.

The current implementation of inequalities finds implicit curves in two dimensions. Generalizing this to three dimensions turns it into a problem of rendering implicit surfaces. This poses an interesting challenge in the design of a protocol to separate the user interface for interaction with graphs from the computation system. To render $f(x, y, z) = 0$, rotate the equation to produce $f'(x', y', z')$ where z' is the axis through the screen. For each screen pixel of fixed (x', y'), find the roots where $f' = 0$ using perhaps an interval extension to Newton's method. The partial derivatives of f give the surface normal (f_x, f_y, f_z) at (x, y, z) where $f = 0$ for the lighting model. The rendering code in the front-end either requires enormous communication bandwidth to the algebra system, or the algebra system needs to provide an executable form of the function f, its derivatives, and an interval extension. Efficiency is of paramount concern in order to render quickly enough to be interactive, making this an interesting problem in protocols. Ray-tracing is embarrassingly parallelizable. (The computation cost per pixel is enormous; the communication cost of the result is negligible.) Thus, this is a perfect problem for parallelizing across many machines, even on a slow network, which further stresses the need for development of protocols between front-ends and potentially many back-ends.

Since the Calculator shipped, I've used it as a code base to explore how symbolic computation can enhance plotting. These changes are described in Avitzur et al. (1995). In two-dimensional graphs, interesting points such as zeros, extrema, and intersections are automatically labelled with symbolic coordinates if a symbolic solution can be found. In three-dimensional plots, interval arithmetic is used to check that polygons aren't drawn spuriously across singularities in the function. Implicit surfaces in three dimensions were also implemented as described above.

Since the release of the product we have heard some encouraging stories about its use in the classroom as a presentation tool. To encourage this, Key Curriculum Press has released *Introducing Dynamic Algebra with NuCalc: Investigating Symbols, Functions, and Graphs* (Erickson 1995) as part of the series *Teaching Ideas for the Mathematics Classroom*. They also distribute the product commercially under the name NuCalc for older 68000-based Macintosh computers.

The most interesting development in user interfaces today is the standardization of compound document architectures at the system level by vendors of system software. Specifically, OpenDoc (OpenDoc Design Team 1994) and OLE 2.0 (Microsoft Corp. 1994) allow applications to embed components which can be edited in place. For example, any word processor which supports these protocols could be used as a notebook environment for a mathematics system which is made of stand-alone components. The need for standardization of communication protocols is obvious.

10 Conclusion

Here we presented the design of a user interface which allows high-school students to explore graphs with neither training nor documentation. We demonstrated that numeric methods, symbolic manipulations, as well as interactions with curves and surfaces can be done without the use of command lines, menus, or dialogs. Direct manipulation is a powerful metaphor for many diverse actions. These design principles can also be applied to mathematics systems designed for expert users. Through progressive disclosure, powerful systems could have default behaviors shielding users from options unless they are desired. By shortening the learning curve, more users will take advantage of computer algebra. The author hopes that groups building mathematics systems will learn from the human interaction community and make usability testing an integral part of interface design.

Acknowledgments

I would like to thank all contributors here, with special mention to Greg Robbins for much of design and implementation, Steve Newman for the 3-D rendering package, and Arnaud Gourdol for extensive design feedback. Pierre Cesarini sponsored the project within Apple. Paula Z. Brown did the graphic design. Michael Tschudy performed the usability study. Chris Forden, Dennis Gately, and Dennis Johnson were responsible for software quality testing. Susan Torres oversaw internationalization. Thanks also to Sonya Andreae, Dylan Ashe, Sam Barone, Mike Bell, Joe Britt, Chris Butler, George Corrick, Tim Dierks, Jeff Elliott, Dave Evans, Richard Fateman, Dave Falkenburg, Winston Hendrickson, Ian Hendry, Carl Hewitt, Shannon Holland, Phil Koch, Ludis Langens, Mike Neil, Andy Nicholas, Alex Rosenburg, Steve Roy, Steven Sargent, Ali Sazegari, John Thompson-Rohrlich, Robert Ulrich, Qarin Van Brink, Mike Wiese, Paul Wolf.

References

Avitzur, R., Bachman, O., Kajler, N. (1995): From honest to intelligent plotting. In: Levelt, A. H. M. (ed.): Proceedings of the ACM International Symposium on Symbolic an Algebraic Computation (ISSAC '95), Montreal, Canada. Association for Computing Machinery, New York, pp. 32–41.

Avitzur, R. (1988) Milo (a Macintosh program). Paracomp Inc., San Francisco, CA.

Avitzur, R. (1989): FrameMath (a software component of FrameMaker). Frame Technology Corp, San Jose, CA.

Bonadio, A. (1989): Theorist (a Macintosh program). Prescience Corp, San Francisco, CA.

Demana, F., Waits, B. K. (1988): Using computer graphing to enhance the teaching and learning of calculus and precalculus mathematics. In: Proceedings of the Conference on Technology in Collegiate Mathematics: The Twilight of the Pencil and Paper, Columbus, OH. Addison-Wesley, Reading, MA, pp. 1–11.

Erickson, T. (1995): Introducing dynamic algebra with NuCalc: investigating symbols, functions, and graphs. Key Curriculum Press, Berkeley, CA.

Fateman, R. (1992): Honest plotting, global extrema and interval arithmetic. In: Wang, P. S. (ed.): Proceedings of the ACM International Symposium on Symbolic and Algebraic Computation (ISSAC '92), Berkeley, California. Association for Computing Machinery, New York, pp. 216–223.

Gourdol, A., Lafourcade, M. (1987): WinMath (a Macintosh program). Winsoft.

Kajler, N., Soiffer, N. (1998): A survey of user interfaces for computer algebra systems. J. Symb. Comput. (to appear)

Microsoft (1994): OLE 2.0 programmer's reference, working with Windows objects. Microsoft Press, Redmond, WA.

Mudur, S. P., Koparkar, P. A. (1984): Interval methods for processing geometric objects. IEEE Comput. Graph. Appl. 4/2: 7–17.

Norman, D. A., Draper, S. W. (eds.) (1986): User centered system design: new perspectives on human–computer interaction. Lawrence Erlbaum Associates, Hillsdale, NJ.

OpenDoc Design Team (1994): OpenDoc, the required reading packet.

Soiffer, N. (1991): The design of a user interface for computer algebra systems. Ph.D. thesis, University of California, Berkeley, CA.

Suffern, K. G., Fackerell, E. D. (1991): Interval methods in computer graphics. Comput. Graph. 15: 331–340.

Tufte, E. R. (1990): Envisioning information. Graphics Press, Cheshire, CT.

Successful pedagogical applications of symbolic computation

Raymond Ravaglia, Theodore Alper, Marianna Rozenfeld, and Patrick Suppes

1 Introduction

At the Education Program for Gifted Youth (EPGY) we have developed a series of stand-alone, multi-media computer-based courses designed to teach advanced students mathematics at the secondary-school and college level. The EPGY course software has been designed to be used in those settings where a regular class cannot be offered, either because of an insufficient number of students to take the course or the absence of a qualified instructor to teach the course. In this way it differs from traditional applications of computers in education, most of which are intended to be used primarily as supplements and in conjunction with a human teacher.

Since the fall of 1990 EPGY has developed a body of computer-based courses designed to teach mathematics from the Kindergarten level through the first two years of university level instruction. Students participating in EPGY enroll formally in these courses through the Stanford University Continuing Studies Program and receive credit for the courses they complete. To date we have had over six hundred students complete courses with us, with about one quarter of those in university level courses. We expect the total enrollment in EPGY for the Fall quarter of 1995 to exceed one thousand students. (For a detailed discussion of our early results, see Ravaglia et al. 1995).

In reading this article it is essential to keep in mind that our software is not something which has remained in a laboratory being poked at by computer scientists or mathematicians, only occasionally to be used by students. Rather students are using it regularly, and as such much of our development has been motivated by their day-to-day needs. Theoretical concerns have often had to assume a secondary role. While this has been frustrating at times, it has also forced us to be extremely honest in evaluating what aspects of our courses play a significant pedagogical role and what aspects are extraneous.

In designing these courses we have tried to be as responsive as possible to variations in student abilities and rates of learning. Because of the opportunities for assessment it affords, symbolic computation has been an essential tool in the instructional design of these courses. In this paper we will briefly examine the standard ways in which symbolic computation is incorporated into mathematics instruction. We will focus not on the benefits that such an incorporation affords, but rather on how specific features common to symbolic-computation programs diminish their pedagogical effectiveness.

These shortcomings will be addressed in the context of a discussion of the EPGY course software in which we will illustrate how we have incorporated symbolic computation into the program and what pedagogical effects it has had. In particular, we will describe how existing symbolic-computation systems suitably modified and coupled with a semantic component, make possible systems in which students can construct derivations and receive immediate feedback. Such derivation systems make it possible to isolate and develop computational skills, without resorting to standard drill and practice techniques. Furthermore, derivation systems, by forcing students to justify explicitly their inferences, make it possible to evaluate student's understanding of a given body of mathematics. This in turn makes it possible to assess students at a deeper level than is usually done in computer-based courses. Ultimately it is the individualized adaptive nature of instruction afforded by the ability to perform this type of assessment which makes it possible to develop successful computer-based courses.

2 Symbolic computation and pedagogy

The last half dozen years have seen the development of a number of symbolic-computation packages available on personal computers. Coincident with the development of these packages have been numerous attempts by educators, computer scientists, and mathematicians to incorporate these powerful tools into secondary-school- and college-level mathematics curricula. A sense of the shear number of these attempts is nicely illustrated by looking through the presentation list of the Seventh Annual International Conference on Technology in Collegiate Mathematics (ICTCM 1995). Over a third of the presentations involved discussions of how symbolic-computation programs have been used to develop instructional material for standard mathematics courses.

The common thrust of the bulk of these efforts has been to provide students with what is essentially a mathematics laboratory environment in which they can explore the properties of certain mathematical objects without having to do any computations themselves.

Depending on the pedagogical goals underlying it, this instructional approach takes several different forms. In its weakest form, students are presented with an unstructured environment in which they are encouraged to experiment freely with some given mathematical object. For instance, students might be given a certain function and told to search for interesting properties that it might have. In the strongest form this approach leads to programs in which the exposition of the curriculum and the symbolic computation package will be linked together. In these cases students will be required to investigate specific properties of mathematical objects (graphs, tables, or sequences of calculations) within the symbolic-computation environment in order to complete an assignment.

Within this model of mathematical experimentation, different features of symbolic-computation programs are emphasized. The most common is to utilize the graphing capabilities of symbolic-computation packages to encourage students to think visually about functions. Perhaps the finest tool developed towards this end is the Graphing Calculator developed by Ron Avitzur (see his contribution in this volume for a thorough discussion of this tool). The influence of this graphical approach is also present in the efforts of calculus revisionists to incorporate graphing calculators into calculus instruction. (A development of calculus which thoroughly incorporates the use of graphing calculators is Hughes-Hallett et al. 1992.) That the merits of this approach are widely accepted is evidenced by the decision of the College Board, a national organization which sets standards on how to teach college-level courses in secondary schools, to require the use of graphing calculators on the national Advanced Placement Examination in Calculus beginning with the 1994–95 exam year.

The other feature of symbolic-computation programs most commonly used is their ability to do infinite precision numerical computations. Such computations are frequently used to illustrate points about functions and limiting processes. By actually evaluating functions at points very close to the limit value, one can obtain very nice numerical tables demonstrating limiting behavior. Similarly one can demonstrate facts about derivatives by computing difference quotients at various points close to the point in question. Because of the ease and rapidity with which one can do these numerical computations, symbolic-computation programs make it feasible to incorporate such examples into the curriculum.

A project which thoroughly incorporates these features of symbolic-computation programs into mathematics instructions is the Calculus and Mathematica Project at the University of Illinois (Brown et al. 1991; for other suggestions, see Crandall 1989, Wagon 1991). Uhl and his colleagues have abandoned the traditional lecture style of teaching calculus for an interactive laboratory setting in which students use Mathematica to work through a series of problems and examples combined with instructional text.

We will not discuss the merits of these approaches here as they are addressed extensively in the literature (again see Brown et al. 1991, Crandall 1989, Wagon 1991). Instead, we wish to focus on specific deficiencies of standard symbolic computation packages when used as pedagogical tools.

2.1 Either too much ...

The greatest weaknesses of standard symbolic-computation programs as pedagogical tools paradoxically stem from their intrinsic power and generality. In their recent book on computer algebra Davenport et al. (1988) state that computer algebra systems as normally conceived are expected to meet two requirements:

- provide pre-programmed commands to perform wearisome calculations;
- provide a programming language to define extensions or enlargements of the original set of pre-programmed commands.

Neither of these desiderata is likely to generate symbolic-computation systems well suited for educational purposes.

The first requirement is problematic because such pre-programmed commands which perform all the "wearisome calculations" obscure many of the details which a student needs to see and learn as part of the process of learning mathematics. Using one command to integrate, for example, ignores the importance such techniques as integration by parts, or integration by trigonometric substitution. The importance of these techniques transcends the merely computational, and teaching them comprises a good deal of the third quarter of calculus. While Maple's "simplify" command might embody all the knowledge of a second-year algebra course, teaching someone how to apply this command in Maple is very different from teaching him second-year algebra.

Also obscured here are many of the subtleties involved in performing certain calculations. The standard symbolic-computation programs do not worry about this sort of error checking since they are designed for use by scientists who already know the mathematics, and can detect erroneous results stemming from incorrect applications of rules and ignore them. This is not an assumption which should be made about someone just learning the material. What is desirable for the expert is often harmful to the student.

The second point is equally problematic. Because programs like Maple and Mathematica have been designed for general use by scientists and engineers, certain assumptions have been made concerning the user interface and the expected power of the system. Scientists and engineers can be assumed to be well versed in programming languages. For this reason, command-language interfaces are the norm for symbolic computation programs. This is understandable since adopting a command language similar to a programming language enables one to create a symbolic-computation environment which is readily extendable.

Students, however, cannot be expected to master anything this complex. While it is reasonable to expect students to use calculators in mathematics courses since the interface of a calculator is almost transparent, it is not a priori reasonable to expect students to learn the interface required to use a symbolic-computation program. Moreover, even if being proficient with the use of symbolic computation programs correlates with an ability to learn calculus, this fact does not justify the introduction of symbolic computation. A similar claim might be made concerning proficiency with first-order logic. It is probably the case that a student who knows first-order logic is not subject, when learning calculus, to the same sorts of confusions concerning quantifiers in proofs using ϵ–δ methods as someone who has not had logic. It does not follow, however, that students should be required to take a course in logic before they take a course in calculus. For this to happen one must first establish that the students who have difficulties with these concepts in the calculus will have an easier time learning them in a logic course. And even then one must show that the time spent first learning logic might not have been more productively spent otherwise.

2.2 ... or too little

The single greatest advantage that computers have as an instructional tool is their ability to individualize instruction to meet the needs of a particular student. This principle, recognized since the earliest drill-and-practice mathematics programs, remains the essential advantage of computer instruction, for while a teacher lecturing must address an entire class at once, a computer, by assessing each student's understanding, and tracking that student's performance over time, can adjust the level of instruction to match that particular student's needs.

Ideally both the amount of material presented to a given student and the level at which it is presented should vary according to the rate at which the student masters the material. Students who can move quickly through a given subject should be allowed to, while those who need additional instruction or a more concrete style of presentation should have those resources available to them. To accomplish this within a computer-based course requires assessment of a student's understanding at each point in the course, together with a record of that student's performance through the course up to that point. The judicious use of symbolic computation makes this type of assessment possible.

It is surprising therefore that it is exactly this type of assessment which is missing in the standard incorporation of symbolic computation into the curriculum. The reasons for this lie in the limitations of the "Mathematica Notebook" approach to course development.

In the notebook model students are presented with what essentially are dynamically unfolding objects. Students can explore these objects and interact with them in real time. One can click on the equation of a function and produce its graph, rotate it in three space, and automatically evaluate its integral. Or one can investigate a function's limits by creating a table of values with arbitrarily great decimal precision. While such experimentation has its place in a computer-based mathematics course, it is not sufficient in itself to constitute a course.

The biggest problem with those who decide to develop courses within the confines of the notebook model is that it is fundamentally a static, non-adaptive model since the notebook cannot modify the presentation of material to reflect the results of student assessment. Consequently, the interaction the students have with these notebooks occurs independently of what their degree of mathematical sophistication is. This is not only a poor utilization of a powerful resource, it is a poor approach to instruction.

3 The EPGY course software

The EPGY course sequence in mathematics consists of on-line courses which cover the standard curriculum in the following courses: First-Year Algebra, Second-Year Algebra, Precalculus (includes Trigonometry and Analytic Geometry), Differential Calculus, Integral Calculus, Multivariate Calculus, Linear Algebra and Differential Equations. Each of these courses is designed to be complete and comparable to what a student would receive in a standard high-school- or college-level course. In this paper we will focus on the sequence from Beginning Algebra through the Integral Calculus.

The EPGY courses are completely computer-based. They are distributed on two or more CD-ROMs and are designed to run locally. Each course consists of lessons which correspond to the logical sections in a textbook. Lessons begin with a multi-media presentation in which digitized sound is played and synchronized in real time with the display of graphics to create what resembles a teacher writing on a blackboard while lecturing. Students have full control over the lectures, being able to pause, fast-forward and rewind at any time. Students can control several lecture parameters including the speed of the lecturer's voice and the format of the graphic display (see Figs. 1 and 2).

It is worth noting that these lectures have been designed so as to preserve the informal nature of spoken mathematics or physics as contrasted with the more formal prose style of textbooks in these subjects. This is important since it has been observed by many people, though never adequately researched, that oral lectures are an important part of learning the mathematical and physical sciences. We believe that such informal lectures are important; they allow the students to absorb matters of style, such as how to talk informally about the subject, how to draw diagrams, and how to write equations.

The lectures are followed by a set of simple questions which review the students' understanding of the material just presented. After these review questions students are presented with a set of interactive exercises, which consist either of a quiz on the material covered in the lecture, interactive exposition in which the student is led through a detailed argument step by step, or a derivation in which the student is asked to obtain the answer

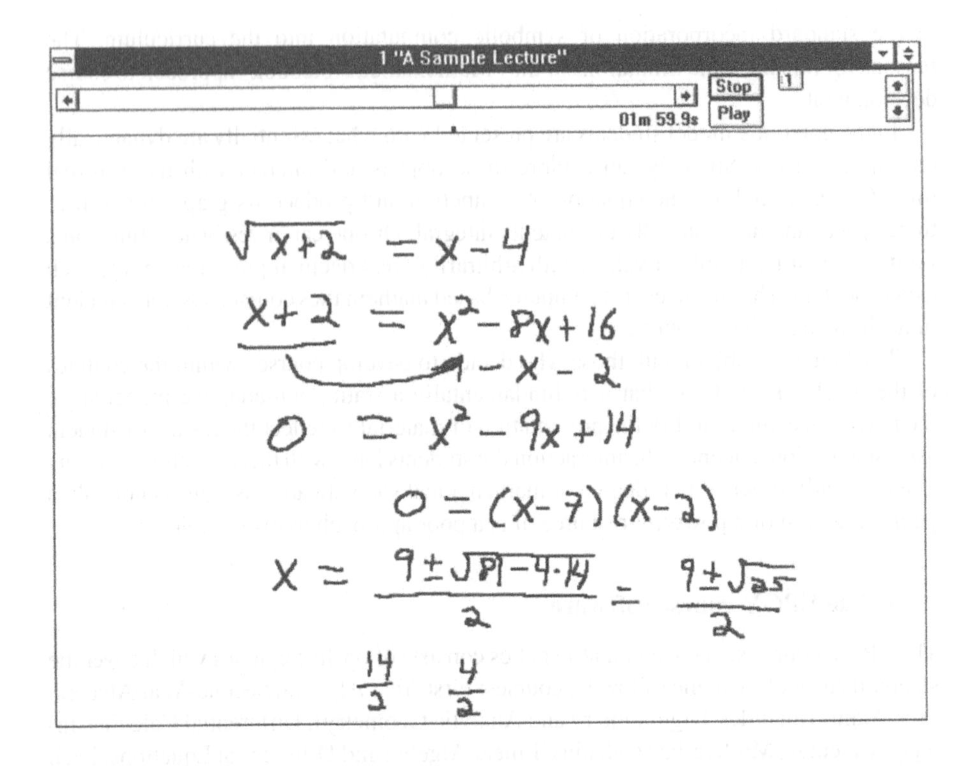

Fig. 1. Handwritten algebra lecture

to an exercise. As one would expect, the difficulty level of the exercises increases as the student progresses into a lesson. Depending on its complexity the student may have to make several intermediary computations. These computations can be done either with paper and pencil or by using our Derivation System.

In addition to the CD-ROM-based instruction students have interaction with remote human instructors and each other through a variety of electronic means. As these features are not directly relevant to the topic at hand, and as they have been discussed elsewhere (Ravaglia et al. 1995, 1994), we will not discuss them here.

3.1 Student assessment and symbolic computation

To be responsive to a student's level of understanding, a computer-based course must be able to assess the student's comprehension of the material and adapt itself accordingly. Students demonstrating ready mastery should be allowed to move quickly through the material while students who are having difficulty should be given appropriate remediation. Ideally this sort of assessment must determine two things. The first is whether or not the student is able to produce the correct answer to the sorts of questions that the student will see on examinations. The second is whether or not the student who produces correct answers actually understands why the answers are correct.

Assessing whether or not students can produce correct answers is relatively straightforward and can be done by asking students the sorts of questions that instructors

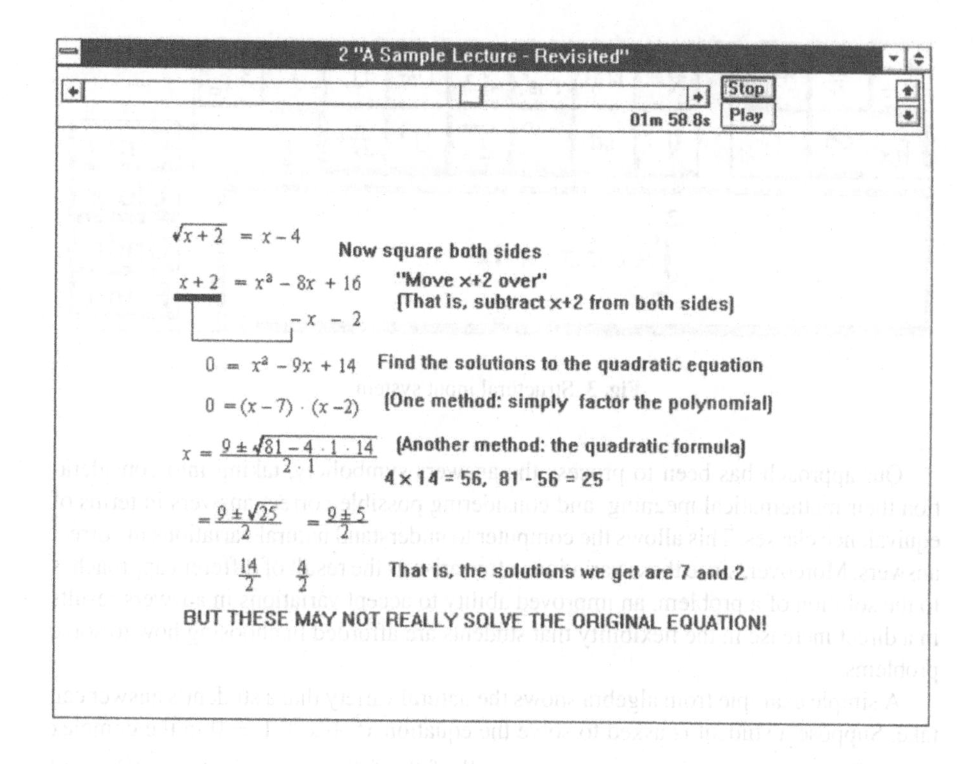

Fig. 2. Formatted algebra lecture

traditionally ask after lectures or on examinations. However, in designing such free-answer questions several issues must be taken into consideration.

The most basic is the ease of input. If a student has to type a complex mathematical expression in an input language, the chance that an incorrect response is caused by an error in typing rather than an error in understanding will be such as to greatly diminish the assessment value of the question. Care must be taken to provide the student with both a convenient means of input such as typesetting programs do, together with the ability to see their input formatted, so that they can verify that what the computer has understood is in fact what they wished to express (see Fig. 3).

Equally important with ease of input is the naturalness of input. Students should not have to constrain their answers to fit a particular form, outside of those constraints which an instructor would reasonably place upon them in class. While it is reasonable to require fractions to be expressed in lowest terms, it is not reasonable to prefer prime notation to Leibniz notation. Students must be allowed to concentrate on getting the problem correct without worrying about expressing it in a form that the computer can understand.

As such it is not surprising that inflexible processing of student answers is a standard complaint with computer-taught courses. Because answer comparison is most often some variation of string matching, students are required to follow rigid input conventions. Unfortunately this eliminates the wide variety of natural variations in the ways people express even relatively simple mathematical expressions.

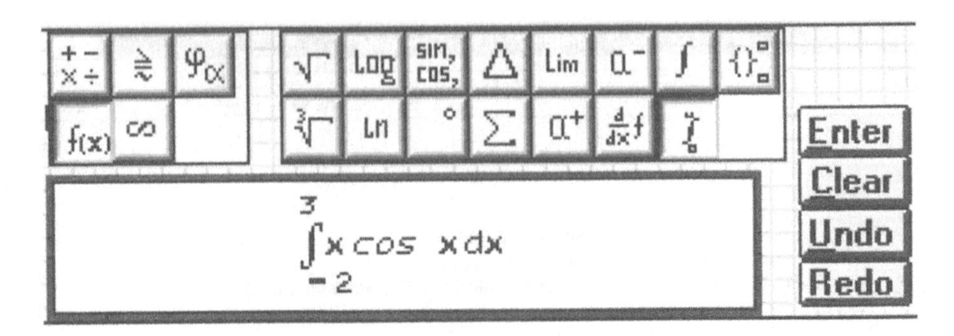

Fig. 3. Structural input system

Our approach has been to process the answers symbolicly, taking into consideration their mathematical meaning, and considering possible correct answers in terms of equivalence classes. This allows the computer to understand natural variations in correct answers. Moreover, since these variations often arise as the result of different approaches to the solution of a problem, an improved ability to accept variations in answers results in a direct increase in the flexibility that students are afforded in choosing how to solve problems.

A simple example from algebra shows the natural variety that a student's answer can take. Suppose a student is asked to solve the equation $x^2 + x + 1 = 0$ in the complex plane. One may want to accept as correct all of the following variants: $\frac{-1+i\sqrt{3}}{2}$ and $\frac{-1-i\sqrt{3}}{2}$; $-\frac{1-i\sqrt{3}}{2}$ and $-\frac{1+i\sqrt{3}}{2}$; $\frac{-1}{2} + \frac{i\sqrt{3}}{2}$ and $\frac{-1}{2} - \frac{i\sqrt{3}}{2}$; $\frac{-1}{2} + i\frac{\sqrt{3}}{2}$ and $\frac{-1}{2} - i\frac{\sqrt{3}}{2}$, not to mention several others with essentially the same form. To code each of these pairs of answers for the purposes of simple string comparison would be a chore and would fail to exploit the semantic content of the mathematical expressions.

Whether or not the student's answer is correct can be determined by passing the student's input and the author-coded answer plus specification of the equivalence class to a symbolic-computation program for evaluation and comparison. Exploiting the fact that the answers are mathematical expressions increases the flexibility for student input and simplifies author coding.

The importance of the author specifying an equivalence class in addition to an answer must be stressed. If full-blown equivalence is tested for between author and student answers, problems in which the student was supposed to have transformed an expression according to algebraic rules into a new expression would be useless. This is because the student could just type in the initial expression, which in the case of problems from an algebra course, is almost always algebraicly equivalent to the answer. This is another example of where the goals of symbolic-computation programs and human instructors diverge.

A good example of this problem are exercises on factoring or multiplying polynomials. The initial expression is algebraically equivalent to the final one, as are all the intermediate stages of the computation, but there is still a definite correct answer. Also, in expressing the factorization of a polynomial, one does not care about differences in order of the terms, or in terms of whether a factor of negative one has been introduced, but one does care that the expression has been factored and as such, one will not extract factors from the student's answer or multiply out the author's. Thus if the problem is to factor $x^2 - 5x + 6$ one would accept as correct $(x-2)(x-3)$ or $(x-3)(x-2)$ or $(2-x)(3-x)$, and so on, but not $x(x-5)+6$.

The size of the set of answer variants which one will accept as correct will depend on the course the student is in as well as the material recently covered in that course. For example a student in beginning algebra, particularly one just learning that i stands for $\sqrt{-1}$, might be required to resolve all negative roots into expressions containing i. In this context $\frac{-1+\sqrt{-3}}{2}$ would not be accepted as equivalent to $\frac{-1+i\sqrt{3}}{2}$, while $-\frac{1-i\sqrt{3}}{2}$ would be. Contrasted to this, an answer from a student in the calculus, just learning derivative rules might be considered correct as long as it was algebraicly equivalent to the author-coded answer and contained no unevaluated functionals.

3.2 The EPGY derivation system

We have just seen how the judicious use of symbolic computation to evaluate answers plays an important role in assessing the ability of students to solve problems. What this use of symbolic computation does not provide in this context, however, is a way to look at the process a student goes through in solving a problem. While getting the correct answer is the goal of solving a problem, getting it in the correct way is equally important. Hence the mathematics teacher's dictum "Show your work."

The EPGY software has addressed this issue by creating a derivation system, i.e., an environment in which students can formally manipulate mathematical expressions by applying inference rules. In the standard environment, the student supplies the rule and the derivation system takes care of performing the appropriate calculation. The results of the calculation are preserved for the student to further manipulate. A derivation of a problem is the set of steps from the statement of the problem to the solution. In addition to the regular exercises students work in the course, they are required to work a certain number in the derivation system.

A derivation system as we conceive of it differs from a raw symbolic-computation environment in two primary ways. The first is in having the logical structure necessary to represent mathematical inference and logical dependency. The second is in providing students with a set of rules to use which are appropriate to both their level of understanding and position within the course.

The importance of the first point can be quickly illustrated by a trivial example of a fallacious inference permissible in most symbolic-computation systems.

Assume: $a = 0$.
Divide both sides by a: $a/a = 0/a$.
Simplify: $1 = 0$.

If one conceives of the intermediate results of one's use of a symbolic-computation system as being unrelated, one cannot discuss inference. Because most symbolic-computation systems are devoid of a semantic component, they impose no structure on the sequence of steps which a student produces in working a computation, and as such, allow students to produce obvious contradictions. What is needed for students is a system which treats the results as a sequence of related lines moving from an initial assumption via accepted inference rules to a valid conclusion, as one does when giving a proof.

The importance of the second point, the need to provide students with rules appropriate to their level of understanding, is one which most symbolic-computation environments have chosen not to pay attention to. This choice is understandable given the population that such symbolic-computation programs are designed for, especially since to do so is essentially to reduce the power of the system. However, while it is understandable, the choice is damaging from a pedagogical perspective. While it may be the case that a physicist using a symbolic-computation package to solve a computationally tedious problem does not need to worry about inferences in moving step to step to the solution, most students learning the material for the first time are not so sure footed. The major difference between students and experts here lies in the assumption that experts can refine the performance of the software with their own knowledge. Experts can be expected to throw out obviously bad results or massage them into a correct form as necessary. It cannot be assumed that students are able to make these same corrections.

Our contention is that, rather than providing powerful commands and extendability, what a symbolic-computation system must provide to properly function as a pedagogically useful derivation systems is a way to represent mathematical inference and logical dependency.

4 User interface and design issues

The EPGY derivation system is present in our sequence of courses from the first year of secondary-school algebra through first year of college calculus. In designing this system we have tried to make its interface as transparent as possible so as to minimize the amount of time students must spend learning its use and maximize the probability that the incorrect application of a rule is due to a conceptual mistake and not a consequence of bad interface design. We have incorporated the structural input system described above into the derivation system interface as well. We have also been motivated by the desire to make it such that all transformations which can be made correspond to an explicit application of a rule. This is essential if every inference is to be explicitly identifiable. We have also required that the system be sensitive to the differences in ability and levels of understanding that the different students using it will have. A feature essential for a calculus student may not always be suitable or even desirable for a student just learning algebra.

In our design we have eschewed both command language interfaces and contextually driven menu interfaces for a number of reasons. Command language interfaces, the norm in commercial symbolic-computation packages like Mathematica or Maple, require class time be spent learning and mastering their syntax, knowledge which may be useless outside of class. Menu interfaces, on the other hand, are more easily mastered, but may be cumbersome to use. If one does not impose a context-based bias on the selection, the user has to wade through a long list of rule possibilities and select the one he or she wants. If one does impose a context-based bias, then the resulting derivations cannot be said to be natural.

What we have settled on for the rules is a set of nested palettes with icons representing the rules. Both the rule groups and the rules themselves are represented by graphical icons depicting the form of the group or rule. To apply a rule, a student first selects the expression or subexpression to which the rule is to be applied, enters any required parameters for the rule into the input window, and then selects the appropriate rule from the rule window. These palettes differ from menus in having all commands immediately ready

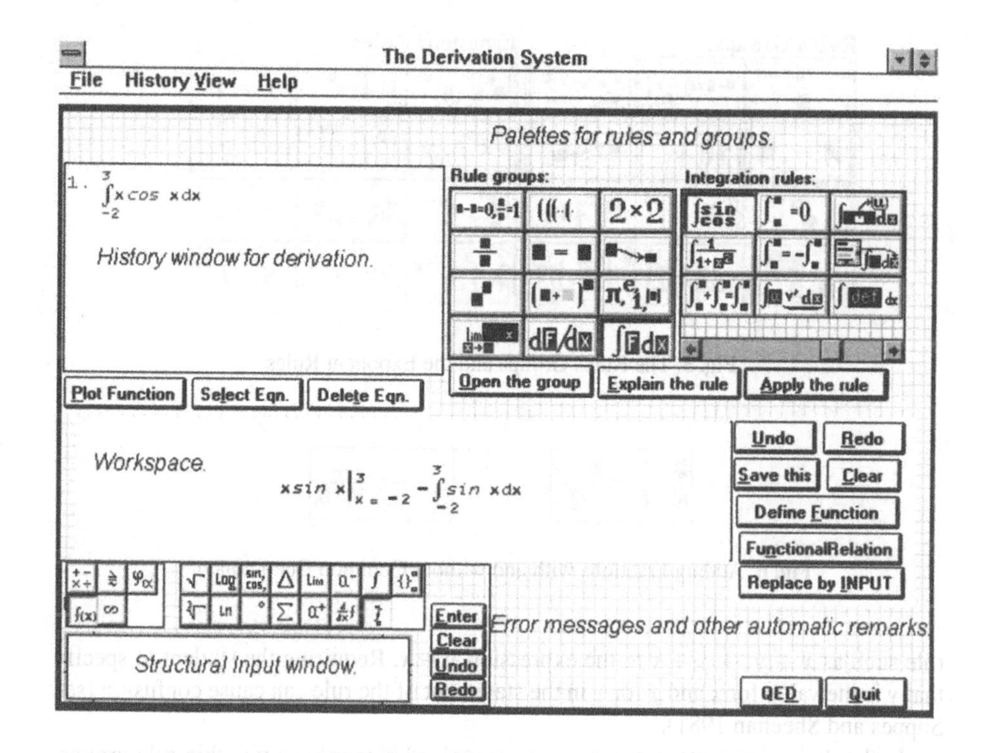

Fig. 4. The derivation system

to hand, rather than buried in linear lists. Moreover palettes naturally lend themselves to the use of icons which in turn exploits the density of information afforded by images.

The icons themselves graphically represent the rules for which they stand, but avoid (in general) the direct use of standard algebraic notation, relying instead on abstract shapes to represent variables or terms within the expressions. So, for example, the exponential rule group is designated by the icon:

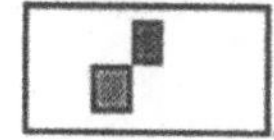.

Selecting the rule group causes the specific rules for that group to appear in the palette to the immediate right of the rule group palette (see Fig. 5).

In the derivation system one can use the mouse to directly select the subexpressions to which one wishes to apply a certain rule. While we have allowed certain keys to cause transformations to the expression being worked on, we have not followed in the direction of Avitzur (see his contribution in this volume) in interpretting mouse movements as commands. The reason for this is the desire to require that students be explicit in their intent to apply rules and in expressing exactly which rule they wish to apply. We believe that exactitude on these issues is important for adequate assessment.

The use of squares with different colors to stand for expressions in the icon, rather than letters, simplifies the process of binding variables when applying rules. This is because the student does not have to suffer the potential confusion caused by having a variable name used in two different ways, such as when a student attempts to apply a

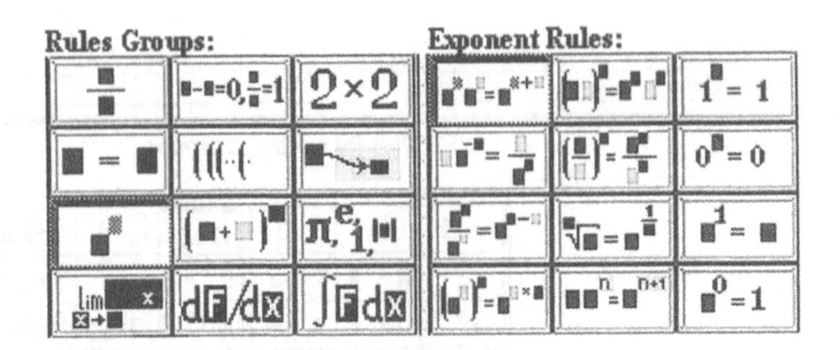

Fig. 5. The Rules Groups and the Exponent Rules

Fig. 6. Adding Fractions with and without Common Denominator

rule such as $x + y \rightarrow y + x$ to the expression $y + x$. Requiring the student to specify that y is the value for x and x for y in the statement of the rule can cause confusion (see Suppes and Sheehan 1981).

In the derivation system, the rules are organized hierarchically within rule groups, according to the type of rule they are. Exponential rules are grouped together, as are rules for grouping, limit rules, differential rules, integral rules, and so on. For convenience we do not adhere to strict categorization, but allow the same rule to belong to several groups. Thus the rule which says that exponentiation by unity leaves an expression unchanged is placed in both the group of exponentiation rules, and in the group of rules reserved for arithmetic simplifications.

This sort of arrangement allows students to find quickly the rule they need to perform a given calculation. If a student wishes to add two fractions, the student first selects the fraction rule, and then determines if the fractions have a common denominator, and then applies the appropriate rule (see Fig. 6).

Naturalness of rules

The rules in the derivation system have been designed to correspond to the theorems one learns in these courses. We have augmented these rules with others which are seemingly redundant, but which facilitate doing certain types of computation. We have tried to make the student's approach to solving problems within the derivation system parallel to the approach the student would take solving the problem with pencil and paper. We have hoped in this way to make work done in the derivation system help the students to develop skills which will carry over to their own work.

Integration of exposition and derivations

We have integrated the derivation system into the development of the standard secondary-school curriculum from the beginning. As soon as students learn an algebraic fact in the course they are taught a corresponding rule in the derivation system. In fact the derivation system has been constructed in such a way that the rules are not present in the system until after the student has formally learned them. In this way the state of the derivation system for a student at a particular time reflects what that student has learned up until that point.

Power of rules appropriate to student understanding

It should be noted that as the students' knowledge increases the power of the rules do as well. Certain computations that should be made explicit by students just learning algebra, would be needlessly tedious to do explicitly by a student in the calculus. The derivation system has been made sensitive to these differences and the rules have been designed to become more automatic in their handling of algebra the further along the student is in the course.

Students must demonstrate comprehension of rules

Every time a rule is introduced into the derivation system, students are expected to spend 5 to 10 minutes working simple problems to familiarize themselves with that rule. Afterwards they work a series of more difficult problems which require use of that rule. In addition to problems which must be solved within the derivation system, students have limited access to the derivation system when working other problems and may choose to solve some of them with the aid of the system.

There is an issue which arises here, however, concerning the nature of solving problems within the derivation system. As in any symbolic-computation environment, the work that the student actually does is rule application. The student does not have to do the computations. As such, just because a student is able to arrive at the correct answer in the correct way, does not guarantee that the student completely understands the process. For instance, a student might know that to find the derivative of an expression, he must first differentiate, then apply derivative rules, and finally collect terms and rewrite the expression, without actually being able to perform the computations on pen and paper needed to arrive at the answer.

We achieve this within the derivation system by requiring students to earn the right to use a rule by demonstrating that they are capable of producing the correct answer themselves. When a student is first learning how to use a rule, he or she is often asked to supply the result after he or she has attempted to apply the rule. The system displays a partial form of the result and asks the student to supply the missing part. As the student becomes familiar with the rule, we decrease the probability that the student will be asked to supply the results of the application. By forcing a student to do this correctly a certain number of times at first, and occasionally throughout the course, we make sure that the student can do the computation embodied in the rule. Moreover by randomly forcing students to do this throughout the course we can make sure that students do not forget how to do these computations.

Metacognition

Having students articulate their reasons for the steps they take in a derivation not only allows for an evaluation of their understanding, it actually deepens it. Each valid rule of computation must be explicitly acknowledged as it is used. In solving a problem students may consider the power and applicability of several rules before choosing the one they consider most appropriate. Through the repetition of this process throughout a course students' comprehension of material is improved, and they are made more likely to retain what they have learned (Cuasay 1992).

5 Logical structure of a derivation

Because the derivation system is supposed to represent inference, the lines in the derivation must be more than a collection of unrelated statements. Some sort of logical structure must be imposed. This structure needs to be more than chain of implication. Consider for instance the solutions to the equation $|x - 3| = 5$. In solving this problem, one first assumes that $x - 3$ is positive and later that $x - 3$ is negative, and solves the resulting equations in each case. Related rates and relative extrema problems afford even more dramatic examples. Even though we are making several assumptions, we are not interested in each of these assumptions being true at once, but only in what each of these assumptions taken in turn allows us to conclude.

We need to make sure that the conclusion arrived at in each strand is correct, but do not need to worry about the compatibility of the different strands, unless we try to combine them in some way. When a rule is applied which causes two strands to be combined, one must verify that there are no contradictory assumptions in the strands. If there are none, the rule can be applied and the new step added. If there is one, then the rule is not applied and the student is warned that there is a problem with applying that rule. As such, the structure of the derivation is graph-like and not simply linear. Each assumption creates a new branch of the graph, and each conclusion is a node, dependent on the assumptions from which it stemmed.

Intermediate steps in derivations and important partial results are stored on numbered lines in the derivation history. Because of space considerations and custom we write each assumptions on its own line, rather than trying to preserve the true tree structure of the derivation. When a given line is selected a system of color codes is used to indicate how the expressions in the derivation depend on each other. Students who wish to may select to see the entire dependency tree of the derivation. We do not display it all the time because we prefer to keep the immediate focus on the mathematics at hand by minimizing the logical machinery that the students need to be aware of.

5.1 Restrictions and logical machinery

For a derivation system to represent inference it is essential that the system be able to keep track of the assumptions that must be made about a given expression before one can apply a particular inference rule to it. These assumptions arise from restrictions on the application of the standard algebraic rules. We return to our example from above.

We started with the assumption:

1. $a = 0$

and attempted to divide both sides by a. The particular inference rule we wished to apply in this case is the conditional statement: if one divides both sides of an equation by the same term, the result will be an equality provided that the term was not equal to 0.

If we attempt to apply this rule to line 1 using the term a, we can recognize that we have a contradiction, and as such can inform the student of this by displaying an appropriate error message. Thus rather than adding the new line:

2. $a/a = 0/a$

the derivation system displays the message "This rule is not applicable since it involves division by zero."

Ideally, one would require students to demonstrate that the antecedent of the rule is satisfied before one would allow them to apply the rule. This is not always practical, however, since one does not always have an explicit value for a given variable.

For this reason we preserve validity by adding the appropriate assumption to the derivation in the form of a restriction. For instance if we had:

1. $a = 0$

and wanted to divide both sides of the equation by some new term b we would add the lines:

2. $b \neq 0$,
3. $a/b = 0/b$.

The reason we can add these two statements as separate lines rather than as just a single line:

2′. If $b \neq 0$, then $a/b = 0/b$

is because we are interpreting the relation between the lines in the same branch of a derivation as being linked by implication and as such there is no difference between the derivation with lines 1–3 and the derivation with lines 1 and 2′. Because line 3 comes from lines 1 and 2, any inference rule we apply to 3 must not have antecedents which contradict 1 or 2.

We call these lines added to the derivation because of their role in the antecedents of inference rules "restrictions." It should be noted that if we were to leave the conditionals intact, the resulting system would be needlessly more complicated in that it would require that the students make the assumptions in the antecedent explicitly and then apply modus ponens. By having the program add the antecedents automatically to the derivation as restrictions, it makes the resulting system much more suitable for students.

5.2 Additional restrictions

The restrictions discussed above are generated by the algebraic form of the expression in question and can be expressed in a natural way in equational form. This makes it natural to add these restrictions to the derivation system automatically. However, there are three other types of restrictions which also must be accounted for if the derivation system is to adequately represent inference.

The first class of restrictions are assumptions about differentiability and continuity. These assumptions cannot always be translated into equational form. In these cases, even if the derivation system cannot establish the truth of an antecedent assumption, the program should still add the antecedent assumption explicitly to the derivation. Thus if in the case of differentiating $f(x) = a^x$, the program could not determine whether or not f was differentiable, it should apply the rule and then add the assumption of differentiability explicitly to the history of the derivation. If an apparent contradiction is obtained at some later point, then this assumption may be identified as the source of the difficulty.

The second class of restrictions are assumptions concerning the domain over which a stated equality holds. Suppose one has the equations $a = 3$ and $a = 5$. If these statements are supposed to hold at a particular point, then we may not allow some step in a derivation to rely upon both. If a is a function which assumes these values at different points, then internally we assign a an independent variable and track these as $a(t_0) = 3$ and $a(t_1) = 5$ to indicate that a attains these value at distinct locations. This assumption will also determine whether a can be considered as a function in terms of t.

The third class of restrictions are those concerning relations of dependency between variables. Often times in the calculus one makes decisions about which variables are independent and which are dependent. These relations determine the applicability of rules such as the implicit function theorem. Similarly, one's ability to convert equations that contain no explicit function notation into statements about functions in which the dependent variable is explicit depends on the assumptions one has made in the course of the derivation. For example, if one has a derivation containing the lines:

p. $x + y = 3$,
r. $3x + y = 5$

one can combine these equations and solve for x and y if one wants. However, one cannot do this if one has already made assumptions about the dependency relations between x and y. If one has previously decided that in both cases y is to be considered a function of x then the result of combining these equations changes the nature of the dependency, as x is not longer a free variable.

5.3 The need for restrictions

Unless one keeps track of the restrictions as they arise in the derivation system, the system will not meet the needs of students. If the result of a derivation is literally false, than students who do not catch the error will be misled and those who do catch it will be confused. Neither case is desirable.

The calculus is littered with examples where the failure of standard symbolic-algebra packages to track such restrictions allow students to wander into nonsense. For instance, if one has the function

$$f(x) = a^x$$

and asks for its derivative, the standard response will be

$$f'(x) = a^x \log a.$$

What the program does not note is that depending on the value of a there are restrictions that must be placed on the domain of the function. Moreover, the domain of the derivative is even more restricted.

Mathematically knowledgeable users of the software understand these domain restrictions and know how to avoid getting into trouble. A student is in no position to do this, particularly if the software itself gives no warning that there is anything unusual about the result. The problem is that the symbolic package has applied the derivative rule for expressions of the form a^x, viz., $(d/dx)\, a^x = a^x \log a$, without examining issues of domain. A student might conclude from this that $(d/dx)(-2)^x = (-2)^x \log -2$. The derivation system avoids this by the use of appropriate restrictions.

A similar problem arises in integration when trying to evaluate definite integrals such as $\int_{-1}^{1} 1/(x^2 - a^2)\, dx$. It is not enough to apply the symbolic rule, since the validity of the application of the rule depends on satisfying an antecedent. In this case the antecedent requires, among other things, that the expression being integrated is defined across the entire interval of integration.

While the integrands algebraic form adds the restriction $x \neq a$, we must also know that since this expression occurs in the context of an integral, $a \notin [-1, 1]$ must be true as well. Unless the derivation system is sensitive to issues of domain which determine whether or not a certain rule is applicable, the system will allow students to produce meaningless results that cannot but add to their confusion.

5.4 Logical knowledge required

In view of the role that restrictions play in the derivation system, the question arises as to how much logical knowledge the students require in order to be able to use the derivation system. Clearly at a minimum they must be aware of what a derivation is and that the lines are logically connected since students who use the derivation system are essentially constructing proofs.

From the students' point of view, however, these are informal proofs, fairly rigorous, but not numbingly explicit. It is not pedagogically justifiable to require students to be aware of all axioms and rules of inference for the predicate calculus in order to work algebra problems. Students should be made aware only of that which is mathematically necessary for their understanding of the content and effective use of the system. Consequently, though the system internally uses much logical machinery to track the interdependencies of the lines of work, students need only have a passing familiarity with the most natural and mathematically essential aspects of logical deductions in order to use the system.

5.5 The underlying engine?

At this point we would like to address a question which arises naturally out of the above critiques of commercial symbolic-computation packages, namely whether or not they have any use at all within a system designed for students. As we stated in the beginning of this paper, we do not dispute the claim that such programs can play important roles in enabling students to explore mathematical concepts. What we wish to examine is whether commercial programs have a place in the role played by our derivation system.

There are two separate questions to be addressed here depending on how wide a role one wishes to assign to the program. If one is asking whether something like Maple is suitable by itself for this role, the answer will have to be a resounding no. For reasons articulated by ourselves throughout this paper (for similar arguments, see the contribution of Beeson in this volume and Kajler and Soiffer 1998), Maple is clearly not up to the task unaided. The question therefore becomes whether a commercial system can have any place within a pedagogically adequate system.

This second question is more subtle. In his paper in this volume, Beeson gives arguments that the answer to this question should also be no. While we appreciate his arguments we do not share in his conclusion. In fact our derivation system as it currently stands makes use of the C-callable kernel of Maple V release 2. This gives us access to the vast knowledge built into the Maple library. We have built our own semantic machinery but we have not developed a full-blown symbolic-computation system from scratch.

The reasons underlying our decision for this are twofold. The first is our recognition that programs like Maple represent massive programming efforts coupled with the feeling that to repeat such an effort would be a waste of resources. It is our belief that, suitably used, such symbolic-computation programs provide a powerful resource.

Exactly how powerful a resource they provide is something that can only be determined after the derivation system has been completed. It may be the case that we will find that unless we have complete control of all computations from start to finish we cannot preserve all the information we need. In this case we will find ourselves at the end of the day having supplied all the features of Maple that we use and our position will have converged with Beeson's.

Our second reason for using Maple stems from the realization that when students complete this course and move into more advanced courses they will find themselves working with existing symbolic-computation programs. To prepare them for this the course must ween them from their dependency on the error checking provided by the derivation system and help them develop their own ability to act as their own auditors. By incorporating Maple into our derivation system and giving them access to it through the course we hope to facilitate their eventual transition to regular users.

6 Two examples

The following two examples illustrate how the derivation system is used. The first example shows a problem from algebra and discusses how students at different levels might solve it. The second example shows a problem from the integral calculus.

6.1 Example 1

Problem. To factor $2x^3 + x^2 - 18x - 9$.

A first-year algebra student

A first-year algebra student encountering a problem like this has just learned about combining like terms, and grouping. Consequently, he or she should attack this problem by grouping terms, to see if common factors exist. In this case, one could highlight $2x^3 + x^2$, then click on the "add parentheses" rule in the *Grouping* rule group. One may then factor out x^2 simply by highlighting the terms in the parentheses, holding the

SHIFT key down while highlighting x^2 (or typing x^2 in the input window), and applying the "Factor out an expression" rule. This yields $x^2(2x+1)$.

Similarly, one groups the $-18x - 9$ terms together, and factors out -9 to yield $x^2(2x + 1) - 9(2x + 1)$. Finally, using either the "Combine like terms" rule or the "Distributive" rule (both in the *Grouping* rule group) with the entire expression highlighted will give $(x^2 - 9)(2x + 1)$. (One could also highlight the whole expression, then highlight just one of the $2x + 1$ terms with the SHIFT key held down, and apply the "Factor out an expression" rule.)

At this point, one may factor $x^2 - 9$ in three ways:

1. by using the "Difference of squares" rule (after replacing 9 by 3^2) or
2. by typing one of its factors in the input window and using the "Factor out an expression" rule;
3. if one is far enough in the course to have received the quadratic formula, one may apply the "Factor quadratic" rule.

If one noticed a different pattern, one could solve the original problem with this same method after first transposing the middle two terms. Highlight the $-18x$ term and hit the left arrow key: the result will be $2x^3 - 18x + x^2 - 9$. One may factor $2x$ out of the first two terms in the way described above, then group the last two terms to get $2x(x^2 - 9) + (x^2 - 9)$ and proceed as before.

A second-year algebra student

A somewhat more sophisticated student might solve the problem by directly dividing out factors. To do this, one highlights the entire expression, types a factor into the input window, and applies the "Factor out an expression" rule.

If one sees that 3 is a root of the polynomial, for instance, one could factor out $x-3$. Selecting the polynomial, typing $x - 3$ into the input window, and applying the "Factor out an expression" rule yields $(x-3)(2x^2+7x+3)$. One could then factor $2x^2+7x+3$ in a variety of ways:

1. If one sees a factor, one may apply the "factor expression" rule again.
2. One may complete the square. First factor out 2 using the "factor expression" rule. This gives $2\left(x^2 + \frac{7}{2}x + \frac{3}{2}\right)$. Then replace 3/2 by $3/2+49/16-49/16$, and use the arrow keys: $2\left(x^2 + \frac{7}{2} + \frac{49}{16} + \frac{3}{2} - \frac{49}{16}\right)$. Use the "Add parentheses" rule to put the perfect square together, then apply the "Perfect square" rule. We now have

$$2\left(\left(x+\frac{7}{4}\right)^2 + \frac{3}{2} - \frac{49}{16}\right).$$

Now highlight the $\frac{3}{2} - \frac{49}{16}$ and use the "Arithmetic" rule. This gives

$$2\left(\left(x+\frac{7}{4}\right)^2 - \frac{25}{16}\right)$$

which can be simplified further by applying the "Difference of squares" rule (after replacing 25/16 by $(5/4)^2$. Now we have

$$2\left(x+\frac{7}{4}+\frac{5}{4}\right)\left(x+\frac{7}{4}-\frac{5}{4}\right)$$

which, after using the "Arithmetic" rule, becomes

$$2(x+3)\left(x+\frac{1}{2}\right).$$

3. One may apply the "Factor quadratic" rule.

A student in a high-level course who needs to factor this polynomial, might use any of the above methods; but he or she would also have access to the "MAPLE factorize" rule, which can perform the factorization in one fell swoop, leaving him or her to focus on the application of the factorization to whatever the real problem at hand might be. Similarly, the "Factor quadratic" rule is usable only by students in second-year algebra, as it applies the quadratic formula. Students at the beginning of the algebra course are not allowed to use even the "Factor out expression" rule until they demonstrate a proficiency with factoring using still more basic rules.

6.2 Example 2

Problem. To evaluate $\int e^{2x}\cos(3x)\,dx$.

Solution. We begin with the definition of a function (line 1):

$$f(x)=e^{2x}\cos(3x).$$

We click on the *Integral* rule group and choose the rule "Antidifferentiate both sides." This gives us the equation:

$$\int f(x)\,dx=\int e^{2x}\cos(3x)\,dx$$

which we save as equation 2 in the history window.

We now choose the "Integration by parts" rule in the *Integral* rule group. We need to highlight the term that will be dv/dx in the integration-by-parts formula $\int u(dv/dx)dx = u\cdot v-\int v(du/dx)dx+C$. If we choose $\cos(3x)$ (by holding down the mouse button and passing over the term) we will get:

$$\int f(x)\,dx=e^{2x}\cdot\frac{1}{3}\sin(3x)-\int\frac{1}{3}\sin(3x)\cdot 2e^{2x}dx+C.$$

We could choose to save this to our history window as is, but we instead do some additional simplification first. Using the arrow keys and "Arithmetic" rule we collect the coefficients in the integral:

$$\int f(x)\,dx=e^{2x}\cdot\frac{1}{3}\sin(3x)-\int\frac{2}{3}\sin(3x)\cdot e^{2x}\,dx+C.$$

And now, the "Extract a constant factor" *Integral* rule gives us:

$$\int f(x)\,\mathrm{d}x = \mathrm{e}^{2x} \cdot \frac{1}{3}\sin(3x) - \frac{2}{3}\int \sin(3x) \cdot \mathrm{e}^{2x}\,\mathrm{d}x + C.$$

Again, we could click on the "save" button, but it isn't really necessary.

We now apply integration by parts again, and again choose the trigonometric function inside the integral (this time it's $\sin(3x)$) to play the role of $\mathrm{d}v/\mathrm{d}x$ (if we choose the exponential function, we will simply undo all our work!). Now we have:

$$\int f(x)\,\mathrm{d}x = \mathrm{e}^{2x} \cdot \frac{1}{3}\sin(3x) - \frac{2}{3}\Bigg(-\mathrm{e}^{2x} \cdot \frac{1}{3}\cos(3x) - \int \left(-\frac{1}{3}\cos(3x) \cdot 2\mathrm{e}^{2x}\right)\mathrm{d}x + C_1\Bigg) + C.$$

This time, we'll save our formula as line 3 in the history window – in case we go wrong in the simplification, we can return to this without having to redo our work. Now we will do more simplification. First, as before, we bring the $-2/3$ out of the integral.

$$\int f(x)\,\mathrm{d}x = \mathrm{e}^{2x} \cdot \frac{1}{3}\sin(3x) - \frac{2}{3}\Bigg(-\mathrm{e}^{2x} \cdot \frac{1}{3}\cos(3x) + \frac{2}{3}\int \cos(3x) \cdot \mathrm{e}^{2x}\,\mathrm{d}x + C_1\Bigg) + C.$$

The integral which remains is the same as the right-hand side of line 2 in the history window: that is, it is equivalent to $\int f(x)\,\mathrm{d}x$! To make the substitution, we must rearrange the terms inside our integral to match the order of the terms in the integral in line 2. This is done with the arrow key. Next, we highlight the whole integral, select line 2 and choose the "Substitute a value for an expression" rule from the *Substitution* rule group. This gives us:

$$\int f(x)\,\mathrm{d}x = \mathrm{e}^{2x} \cdot \frac{1}{3}\sin(3x) - \frac{2}{3}\Bigg(-\mathrm{e}^{2x} \cdot \frac{1}{3}\cos(3x) + \frac{2}{3}\int f(x)\,\mathrm{d}x + C_1\Bigg) + C.$$

We save this as line 4 and now need only do some basic algebra to finish. We use the distributive rule (in the *Grouping* rule group) to multiply out $-2/3(-\mathrm{e}^{2x} \cdot 13\cos(3x) + 2/3\int f(x)\,\mathrm{d}x + C_1)$. We now have:

$$\int f(x)\,\mathrm{d}x = \mathrm{e}^{2x} \cdot \frac{1}{3}\sin(3x) + \frac{2}{3}\mathrm{e}^{2x} \cdot \frac{1}{3}\cos(3x) - \frac{2}{3} \cdot \frac{2}{3}\int f(x)\,\mathrm{d}x - \frac{2}{3}C_1 + C.$$

Then, we highlight the $-\frac{2}{3} \cdot \frac{2}{3}\int f(x)\,\mathrm{d}x$ and use the arrow keys to move it to the left-hand side of the equals sign:

$$\int f(x)\,\mathrm{d}x - \left(-\frac{2}{3}\cdot\frac{2}{3}\int f(x)\,\mathrm{d}x\right) = \mathrm{e}^{2x}\cdot\frac{1}{3}\sin(3x) + \frac{2}{3}\mathrm{e}^{2x}\cdot\frac{1}{3}\cos(3x) - \frac{2}{3}C_1 + C.$$

Again, we turn to the *Grouping* rule group. The "Cancel minus signs" rule will bring us to:

$$\int f(x)\,\mathrm{d}x + \frac{2}{3}\cdot\frac{2}{3}\int f(x)\,\mathrm{d}x = \mathrm{e}^{2x}\cdot\frac{1}{3}\sin(3x) + \frac{2}{3}\mathrm{e}^{2x}\cdot\frac{1}{3}\cos(3x) - \frac{2}{3}C_1 + C.$$

And the distributive rule, applied first to the entire left-hand side, then to the first two terms of the right-hand side, will give us:

$$\int f(x)\,\mathrm{d}x\cdot(1+\frac{2}{3}\cdot\frac{2}{3}) = \mathrm{e}^{2x}\cdot\frac{1}{3}\left(\sin(3x)+\frac{2}{3}\cos(3x)\right) - \frac{2}{3}C_1 + C.$$

We may use the "Arithmetic" rule to replace $(1 + (2/3)\cdot(2/3))$ by 13/9; the *Equation* rule group gives us a rule to divide both sides by 13/9; the *Fraction* rule group gives us rules to cancel and simplify terms in the fractions to yield at last

$$\int f(x)\,\mathrm{d}x = \frac{3}{13}\mathrm{e}^{2x}\cdot\left(\sin(3x)+\frac{2}{3}\cos(3x)\right) - \frac{6}{13}C_1 + \frac{9}{13}C.$$

We click on the QED button and are told:

Well Done. You have correctly solved the problem.

There are several things to notice about the derivation above.

The system ignores the constants of integration in determining the correctness of the answer. Doing the problem a different way, say by letting e^{2x} be $\mathrm{d}v/\mathrm{d}x$ for both applications of integration by parts, might well result in different constants of integration.

There is a great deal of flexibility in which lines are saved in the history window. It was necessary to use one early line later in the proof for substitution, but the other lines were saved only for convenience (in case one wanted to start over from a certain stage in the problem) or clarity.

Much algebraic simplification is optional. For this problem, any algebraic answer whose derivative is $\mathrm{e}^{2x}\cos(3x)$ will be accepted. Most simplification, as in off-line work, is for the benefit of the student's vision. Of course, some simplification was necessary to justify the substitution of the original integral back into the equation.

7 What is gained

Having described the derivation system and explained how it addresses several deficiencies found in common symbolic-computation packages, we turn to an examination of the pedagogical benefits derived from it. These benefits, coupled with the semantic processing of student input discussed above, are the true benefits afforded by symbolic computation.

7.1 Instant feedback with flexibility

The most immediate benefit of incorporating the derivation system into a mathematics course is that it allows students to get instant feedback on their work. This is true both with free-response questions solved by students without use of the derivation system, and with answers produced in the derivation system. In traditional courses, several days may pass before a student receives graded work back from an instructor, which may allow erroneous patterns of reasoning to become settled. With immediate answer checking – which extends to checking each step for problems done entirely in the derivation system – students can recognize their misunderstandings as soon as they develop and seek help to correct them.

Instant feedback in computer-based courses does not absolutely require the use of symbolic-computation software; however, such software allows semantic-based checking of answers and maximum flexibility of response. In a similar way, step-by-step evaluation does not absolutely require the use of a derivation system. Some courses try to evaluate students on a step-by-step level by specifying the steps to be applied and prompting students to type in each intermediate result. In essence, this decomposes complex problems into simple problems, each of which must be answered by the student. Yet there may be many correct approaches to solving a complex problem, and many reasonable choices for intermediate results. That students will take wildly different paths in reaching the goal in a derivation has been well documented in both our logic and set theory courses (Kane 1981, Suppes 1981). A derivation system allows step-by-step evaluation while allowing students to decide for themselves what steps to take in solving a problem.

7.2 Forced demonstration of understanding

In addition to focusing on what has been recently learned and in giving students instant feedback, the derivation system plays a crucial role in moving beyond the traditional limitations in evaluating student understanding. Traditional computer instruction evaluates a student's performance solely in terms of whether he or she produces correct answers. It does not examine the process that the student went through to produce those answers. As such, it cannot adequately assess student understanding. A student, especially in mathematics, should always be able to justify how an answer was arrived at.

Forcing students to solve problems in the derivation system by applying rules is one way to evaluate whether or not the students understand the process of how to arrive at the correct answer. If the correct answer is given and the steps taken en route to the answer are all justified, then the student in some sense understands how to produce the answer. This type of evaluation of understanding is what exercises in derivation systems normally afford.

More than this is desirable, however, since within a derivation system arriving at answers does not entail that one completely understands the computational process required to produce the answer. This concern is particularly urgent in courses like algebra, in which one major objective (though by no means the only one) is to teach students how to perform a certain class of computation. To some degree the measure of ones knowledge of algebra is simply how well one can do algebraic computations. While this is certainly not the only criterion for understanding, it is often the sole criterion used on standardized examinations. It is essential to make sure that students not only know which rules might be appropriate in a particular situation, but also are able to apply the transformations themselves.

As mentioned above we achieve this within the derivation system by requiring students to supply partial result after they have applied rules. The system displays a partial form of the result and asks the students to supply the missing part. By forcing a student to do this frequently at first and periodically throughout the course, we make sure that the student can do the computation embodied in the rule.

8 Limitations and desiderata

The EPGY software is constantly under review, as feedback from our students pours in daily. To support our existing student base, we attempt to improve our courses and address observed deficiencies as quickly as possible. At the realities of managing hundreds of students of varying degrees of technical sophistication, each running our software on his or her own computer with its own particular idiosyncrasies make the potential consequences of implementing hastily improvised solutions quite unpleasant. As the number of students using our software has grown we have come to learn the importance of the careful collection and analysis of data concerning how students actually use the program. The importance of usability testing discussed by Avitzur in his contribution to this volume is a lesson we have learned well and often over the past few years.

There are many improvements we hope to add to the derivation system in the next few years. Most immediately, we intend to improve the range and quality of error messages, offering students who attempt to misapply rules a more complete explanation of the nature of their error, together with suggestions on what to do instead. Expanding our existing help system to contain pointers to appropriate places in the course would also help achieve this goal.

Along these lines we would also like to build student proofs directly into the derivation system. At the moment students who wish to see a complete proof of a derivation must contact their instructors. One way, requiring little programming effort but a great deal of editing, would be to hard-wire solutions to each assigned problem in each course; another way would be to extend the system to generate automated solutions to wide classes of problems.

Still another method, intriguing in its use of student data, would be to collect and compare solutions already generated by students running the course and select the best (measured by some automated criteria) to be included in future releases of the course. Incentives could be offered to inspire students to come up with the shortest, or most elegant solution, but the key point is that, as this data is already being collected and analyzed, it would be a natural to automatically incorporate such proofs into subsequent versions of the program.

There are improvements we would like to make involving speech recognition and handwriting input which we mention here without elaboration.

8.1 Validity vs. vacuity

As mentioned above we ensure the literal validity of the derivation by including all constraints and dependencies needed for the valid application of each rule used as additional lines of the derivation. In this way, any finished derivation must be literally correct, as assumptions sufficient for each step are included as hypotheses in the history window.

In doing this one faces a trade-off between halting a student derivation when one detects that the application of a rule leads to a contradiction and allowing the derivation to continue as true but vacuous. While the ideal is to detect errors when possible, we cannot catch all such errors. It is possible to derive a seemingly false result; however, if one carefully checks the restrictions listed in such a derivation one will discover that the domain for which the result holds is empty. The result is logically valid, but may be misleading.

This difficulty is one common to any effort to guarantee the validity of student derivations. Indeed the work done by Beeson faces a similar problem, for as many clerical restrictions in Mathpert are displayed only when requested, it may be possible to derive a seemingly false result *without* seeing the internally stored restrictions that invalidate it.

As we collect data on what sorts of contradictions arise most frequently we hope to improve our ability to detect them when they arise. Even if one cannot in principle notice all contradictions, one may be able to detect most of the ones which arise naturally in the specific body of curriculum for which our derivation system has been designed.

8.2 Forced justification of results

A more extreme way to test the ability of students to both compute and to justify their answers is to transform the derivation system into what we have dubbed a "justification system." A justification system inverts the normal order of the derivation system in requiring students to first perform a computation themselves, and then to justify the computation by selecting the appropriate derivation system rule as justification. The derivation system is then used to evaluate the justification to make sure that the computation performed by the student was correct. This style of work is reminiscent of the traditional two-column style of giving geometry proofs where students are asked to list their statements in the left column and to give the reason justifying the statement in the right column.

While a system like this would be tedious to use if one had to input extensive mathematical expressions with a keyboard, it would not be so if one could input answers with a pen and tablet, using full handwriting recognition of two-dimensional mathematical notation. With this method of input, working in the justification system would be essentially the same as working on paper the way students normally do when forced to show all of their work. By forcing the students to cite the rule which licenses their inference, the derivation system could verify that the student's computation was in fact correct. This would provide the students with the ideal environment in which to do assignments. They would be allowed to progress until they made an error at which point they would be asked to explain themselves. When correct they would be allowed to continue but when incorrect they would arrive at the source of their error. This has definite advantages over the traditional model for student homework and instructor response.

8.3 Automated solution generation

Another desideratum that we have yet to investigate is the automatic generation of solutions to derivations that are appropriate to student levels of understanding. Some promising work has been done by those working in the field of artificial intelligence in education, particular those working on intelligent tutoring systems (Nicaud 1994, Beeson 1990, Oliver and Zukerman 1991). The common approaches to these sorts of systems, combine symbolic-computation environments with models of student problem solving to create tutorial systems in which students can work problems and receive instruction in the form of hints when they are stuck. They can also have the computer solve problems from some set of types.

In the introduction of Nicaud (1994), the author laments the fact that despite much work on intelligent tutoring systems since their inception in 1972, few are currently in use and few have been shown to have good teaching capability. He writes: "Many researchers who worked in this field have changed their goals and are now working on educative but non-teaching environments . . . Many researchers have more or less reached the conclusion that intelligent tutoring systems are *impossible.*" We at EPGY have not reached this negative conclusion. We have concluded, however, that prior to working on such systems one should accomplish two goals, namely:

- the development of a derivation system that allows students to solve problems in a natural way, without causing deviation from the strategies that they would apply when working with pen and paper; and
- the collection of a significant amount of data on how students actually reason about problems, particularly when required to show all their work.

As we deepen our understanding of student solutions we will begin to address this issue.

9 Final remarks

In this article we have discussed the role symbolic computation plays in an existing corpus of computer-based courses. While little in our work may be new from a theoretical standpoint, we have created what we believe to be the largest body of self-contained computer-based courses in mathematics at the secondary-school level that are in general use. This in turn has enabled us to test the effectiveness of new techniques by seeing if they do in fact improve student results. As our courses are refined and we begin to analyze the data we have collected, we hope to deepen our understanding of what just exactly are the pedagogical ramifications of symbolic computation.

The software discussed herein is not commercially distributed. It is solely for use by students taking courses through the Stanford University Continuing Studies Program from the Education Program for Gifted Youth.

References

Ager, T., Ravaglia, R., Dooley, S. (1989): Representation of inference in computer algebra systems with applications to intelligent tutoring. In: Kaltofen, E., Watt, S. (eds.): Computers and mathematics. Springer, New York Berlin Heidelberg, pp. 215–227.

Beeson, M. (1990): Mathpert: a computerized environment for learning algebra, trig, and calculus. J. Artif. Intell. Educ. 2: 1–11.

Brown, D., Porta, H., Uhl, J. (1991): Calculus and Mathematica: a laboratory course for learning by doing. In: Leinbach, L. C., Hindhausen, J. R., Ostebee, A. M., Senechal, L. J., Small, D. B. (eds.): The laboratory approach to teaching calculus. The Mathematical Association of America, Washington, DC (MAA Notes, vol. 20).

Chuaqui, R., Suppes, P. (1990): An equational deductive system for the differential and integral calculus. In: Martin-Löf, P., Mints, G. (eds.): COLOG-88. Springer, Berlin Heidelberg New York Tokyo, pp. 25–49 (Lecture notes in computer science, vol. 417).

Crandall, R. E. (1989): Mathematica for the sciences. Addison-Wesley, Menlo Park, CA.

Cuasay, P. (1992): Cognitive factors in academic achievement. Higher Ed. Ext. Serv. Rev. 3/3.

Davenport, J., Siret, Y., Tournier, E. (1988): Computer algebra: systems and algorithms for algebraic computation. Academic Press, London.

Hughes-Hallett, D., et al. (1992): Calculus. Wiley, New York.

ICTCM (1995): Eighth International Conference on Technology in Collegiate Mathematics, November 1995, Houston, TX, prelimary schedule.

Kaltofen, E., Watt, S. (eds.) (1989): Computers and mathematics. Springer, New York Berlin Heidelberg.

Kane, M. T. (1981): The diversity in samples of student proofs as a function of problem characteristics: the 1970 Stanford CAI logic curriculum. In: Suppes, P. (ed.): University-level computer-assisted instruction at Stanford: 1968–1980. Inst. Math. Stu. Soc. Sci., Stanford University, Stanford, CA, pp. 251–276.

Kajler, N., Soiffer, N. (1998): A survey of user interfaces for computer algebra systems. J. Symb. Comp. (to appear)

Moloney, J. M. (1981): An investigation of college-student performance on the 1970 Stanford CAI curriculum. In: Suppes, P. (ed.): University-level computer-assisted instruction at Stanford: 1968–1980. Inst. Math. Stu. Soc. Sci., Stanford University, Stanford, CA, pp. 277–300.

Nicaud, J. F. (1992): Reference network: a genetic model for intelligent tutoring systems. In: Frasson, C., Gauthier, G., McCalla, G. I. (eds.): Intelligent tutoring systems. Springer, Berlin Heidelberg New York Tokyo, pp. 351–359 (Lecture notes in computer science, vol. 608).

Nicaud, J. F. (1994): Building ITSs to be used: lessons learned from the APLUSIX project. In: Lewis, R., Mendelsohn, P. (eds.): Lessons from learning. North-Holland, Amsterdam, pp. 181–198 (IFIP transactions, series A, vol. 46).

Oliver, J., Zukerman, I. (1991): DISSOLVE: an algebra expert for an intelligent tutoring system. In: Lewis, R., Otsuki, S. (eds.): Proceedings of Advanced Research on Computers in Education, IFIP TC3 International Conference, Tokyo, Japan, 1990. North-Holland, Amsterdam, pp. 219–224.

Ravaglia, R. (1995): Design issues in a stand alone multimedia computer-based mathematics curriculum. In: Fourth Annual x. Multimedia in Education and Industry, Asheville, NC, pp. 49–52.

Ravaglia, R., de Barros, J. A., Suppes, P. (1994): Computer-based advanced placement physics for gifted students. Comput. Phys. 9: 380–386.

Ravaglia, R., Suppes, P., Stillinger, C., Alper, T. (1995): Computer-based mathematics and physics for gifted students. Gifted Child Q. 39: 7–13.

Richardson, D. (1968): Some undecidable problems involving elementary functions of a real variable. J. Symb. Logic 33: 515–521.

Suppes, P. (ed.) (1981): University-level computer-assisted instruction at Stanford: 1968–1980. Inst. Math. Stu. Soc. Sci., Stanford University, Stanford, CA.

Suppes, P., Sheehan, J. (1981): CAI course in axiomatic set theory. In: Suppes, P. (ed.): University-level computer-assisted instruction at Stanford: 1968–1980. Inst. Math. Stu. Soc. Sci., Stanford University, Stanford, CA, pp. 3–80.

Suppes, P., Sheehan, J. (1981): CAI course in logic. In: Suppes, P. (ed.): University-level computer-assisted instruction at Stanford: 1968–1980. Inst. Math. Stu. Soc. Sci., Stanford University, Stanford, CA, pp. 193–226.

Wagon, S. (1991): Mathematica in action. Freeman, San Francisco.

Algorithm animation with Agat

Olivier Arsac, Stéphane Dalmas, and Marc Gaëtano

1 Introduction

Algorithm animation is a powerful tool for exploring a program's behavior. It is used in various areas of computer science, such as teaching (Rasala et al. 1994), design and analysis of algorithms (Bentley and Kernighan 1991), performance tuning (Duisberg 1986). Algorithm animation systems provide a form of program visualization that deals with dynamic graphical displays of a program's operations. They offer many facilities for users to view and interact with an animated display of an algorithm, by providing ways to control through multiple views the data given to algorithms and their execution.

Perhaps the most natural application of algorithm animation is teaching. There is currently a great interest in applying algorithm animation for computer-aided instruction that led to the building of several new systems (Gloor 1992, Rasala et al. 1994).

Another very promising use of algorithm animation could be the understanding, debugging, and tuning of complex algorithms. In computer algebra as in many applications, algorithms are often complex, manipulating sophisticated data structures and exhibiting a complicated behavior at run-time that is very dependent on the input data. It is at least as important to observe a program's progress as it is to obtain a final result, if indeed there is any. For instance, the performance of algorithms to compute the Gröbner basis of an ideal of polynomials is sensitive to the ordering of variables. Animating the computation of such algorithms could indicate heuristics for deducing a good ordering from a given input or demonstrate some subtle bugs or performance problems.

The common approach to animating algorithms specified in high-level procedural languages was pioneered in BALSA in the early 1980s (Brown 1988). Briefly, this approach is as follows: the program to be animated is annotated with markers that select the fundamental operations that are to be displayed. These annotations are called *interesting events* and can have parameters whose values are program data. *Views* are used to give a visual representation of events. Most of the views are graphical and each view controls a *screen*. A view is responsible for updating its graphical display appropriately based on the event. Views can also propagate information from the user back to the algorithm.

Systems such as BALSA were constrained by a lack of computational power for real-time two-dimensional graphics. In the mid-1980s, systems like Animus (Duisberg 1986) used smooth transformations of 2-D images to animate small examples. Tango in the late 1980s (Stasko 1990), and then XTango (Stasko 1992) provided an elegant framework for specifying 2-D animations. XTango is designed to allow users to easily animate an algorithm without having to write any low-level graphics code. This package is an implementation of the *path-transition* algorithm animation framework developed

I will show in this paper, among other things, that if we start with an *educational* purpose, and enunciate some simple design principles that more or less obviously follow from that purpose, these principles have ramifications that run through to the computational core of such a system, so that it is impossible to achieve ideal results by tacking on some additional "interface" features to a previously existing computation system. To put the matter another way: it is not possible entirely to separate "interface" considerations from "kernel" considerations. Such a separation might be possible (to facilitate running on different platforms), but only if the kernel itself has been designed with education in mind.

The most serious and thorough attempt to build an educational system based on pre-existing symbolic-computation software is described in this volume (see the contribution by Ravaglia et al.). The first paragraph of that paper says "we will demonstrate how specific features common to symbolic-computation programs diminish their pedagogical effectiveness." Nevertheless, they say, "The reason to adopt an existing package rather than develop ones own is the recognition that programs like Maple and Mathematica represent massive programming efforts that do not need to be repeated." They identify many of the same problems with computation systems that are discussed in this paper. In order to overcome the problem of incorrect inferences, they developed a separate "derivation system" of "semantic machinery", which runs between Maple and their user interface. Also, a good deal of programming in the Maple language was necessary to produce student-sized steps, rather than mystifying answers. A more detailed comparision of their system with Mathpert is beyond the scope of this paper. Their work does not constitute a counterexample to the claim in the text, but rather additional support: rather than "tacking on" an interface to Maple, it was found necessary to perform many person-years of system development in both C and Maple to obtain a usable system.

Although Mathpert is not designed to replace teachers and books, there is a demand for stand-alone software that would allow a student to work alone, independent of teachers and books. Therefore Mathpert has been designed in such a way that it can later be incorporated in or used by systems which *are* explicitly tutorial. This point will be taken up in the last section of this paper.

3 Design principles

In this section I present and discuss eight fundamental design principles that guided the design of Mathpert.

3.1 Cognitive fidelity

We call mathematical software *cognitively faithful* if it satisfies the following criterion. When generating solutions, as opposed to supporting student-generated solutions, the software solves problems as the student should. This means that it takes the same steps as the student should take, in a correct order. Of course, there may be several acceptable "solution paths", all of which will be accepted if the student takes them; but the one which the computer generates if requested to do so must be exemplary in every respect. In particular, high-powered algorithms based on advanced methods are problematic. Most modern computer algebra systems use such advanced methods for fundamental operations such as factoring polynomials, solving equations, and computing integrals. What is a student to think when she asks for a factorization of $x^5 + x + 1$ and receives the answer $(x^2 + x + 1)(x^3 - x^2 + 1)$?

She will have no idea how such a factorization can be obtained, even though the result can be verified by multiplying out. For a student who has mastered elementary algebra, it might be quite a valuable aid to genuine mathematical exploration to have such an "oracle" available. But, the computer needs to present elementary solutions to elementary problems when the student is learning elementary methods, and it needs to *not* present answers achieved by methods that the student cannot understand.

3.2 Glass box

This means that you can see how the computer solves the problem. The program presents, or allows the student to construct, step-by-step solutions, not just answers. This is a very important point, and often misunderstood. It is comparatively easy for a program to get the one-line *answer* to a problem. It is more difficult to generate a multi-line *solution* with understandable steps. When combined with the requirement that the steps be cognitively faithful, this turned out to be not only the most important but also the most difficult criterion to satisfy. The glass-box criterion requires it to be apparent how each step was obtained; in practice this means accompanying each step with a "justification".

Figure 1 exhibits a sample step-by-step solution, captured from a Mathpert screen.

The simplest computer algorithms almost never generate cognitively faithful solutions. In applications where only the answer is of interest, and not the steps by which it was obtained, this doesn't matter. But in building software for mathematics education, it is crucial. "Human" solutions vary greatly with the features of each particular problem, delicately adjusting the choice and order of operations to be applied so that the resulting solution is economical and beautiful. Software for mathematics education will also have to generate economical and beautiful solutions.

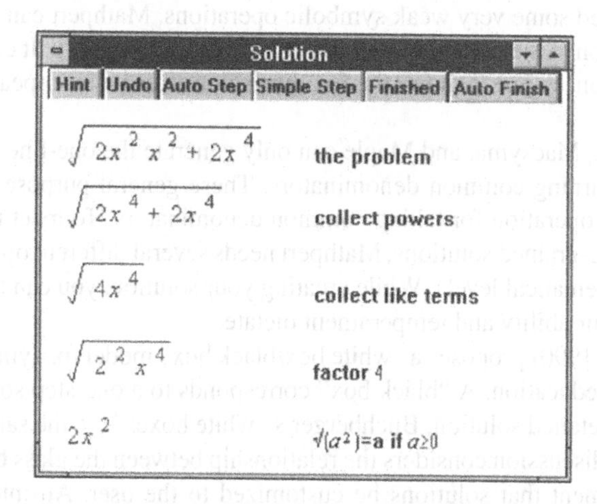

Fig. 1. A sample step-by-step solution

In practice the ability of modern algorithms to solve problems the student cannot solve is not as large an obstacle to the use of existing computer algebra systems as is the inability to break the computation into comprehensible steps. Even when a solution can be obtained by elementary means, it is important to break it into steps. For example, the following factorization of $x^{100} - 1$ can be obtained by a fairly short series of elementary steps, yet can be overwhelming if presented all at once:

$$(x-1)(x^4+x^3+x^2+x+1)(x^{20}+x^{15}+x^{10}+x^5+1)$$
$$(x+1)(1-x+x^2-x^3+x^4)(1-x^5+x^{10}-x^{15}+x^{20})$$
$$(1+x^2)(x^8-x^6+x^4-x^2+1)(x^{40}-x^{30}+x^{20}-x^{10}+1).$$

But if the computation system does not internally factor the polynomial by a series of elementary steps, but instead by a high-powered method such as Berlekamp's algorithm, there is no hope of getting it to generate the required series of elementary steps.

3.3 Customized to the level of the user

Cognitive fidelity demands that the software generate solutions that will be exemplary for the student. But students at different levels need different solutions.

A beginning algebra student needs a five-line solution to the common-denominator problem $1/x + 1/y$. A calculus student evaluating a complicated integral does not want to see five lines devoted to $1/x + 1/y = (x + y)/xy$. It follows that the program must contain some, albeit rudimentary, representations of its user's knowledge. This representation must be sufficient to enable the generation of solutions with many small steps or few, powerful steps, depending on the requirements of the individual student.

It is apparently simpler to allow the user to create long ("fine-grained") or short ("coarse-grained") solutions, than to require the computer to do it. At least a "user model" would not be required. But there are still some difficulties even with user-generated solutions. Namely, this requirement means that the program must have a varied arsenal of mathematical operations. In order to allow the creation of fine-grained solutions, we need some very weak symbolic operations. Mathpert can generate a five-or-six line solution to the common-denominator problem $1/x+1/y$. It can also generate a one-line solution. Figure 2 shows the five-line solution as it can appear on Mathpert's screen.

Mathematica, Macsyma, and Maple can only generate the one-line solution, which is useless for learning common denominators. These general-purpose programs need only one simple operation for taking common denominators. To meet the requirement of generating fine-grained solutions, Mathpert needs several different operations, for use at different mathematical levels. While creating your solution, you can take big steps or little ones, as your ability and temperament dictate.

Buchberger (1990) proposes a "white box/black box" model for symbolic-computation software in education. A "black box" corresponds to a one-step solution, a "white box" to a fully detailed solution. Buchberger's "white boxes" are the same as my "glass boxes". But his discussion considers the relationship between the glass box requirement and the requirement that solutions be customized to the user. An intermediate case, where the solution is in several steps but not at the finest level of detail, could be

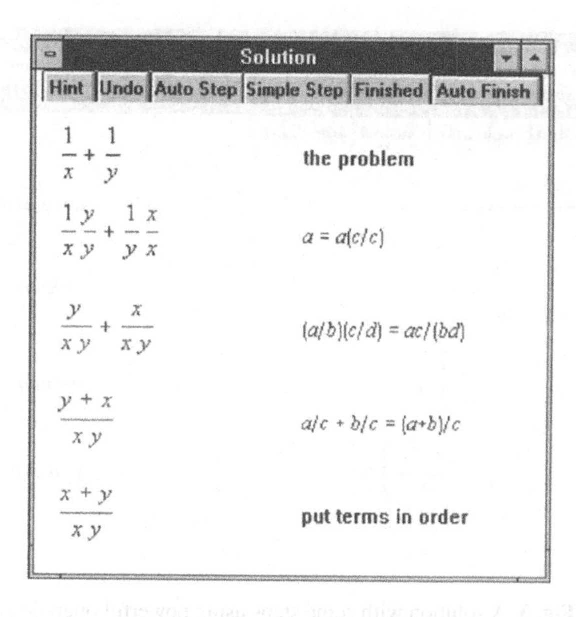

Fig. 2. A solution with many small steps

analyzed in Buchberger's terms as a white box at the top level, in which subordinate steps are treated as black boxes. For example, when integrating rational functions, partial-fraction decomposition can be treated as a black box. Mathpert supports this paradigm fully, allowing the user complete flexibility to customize the level of detail desired in the solutions. It is a delight, when working out a step-by-step solution to an integration problem, to just push a button to get a partial-fraction decomposition. Figure 3 shows the beginning of a solution to an integration problem, in which two complicated algebraic steps are taken at a single mouse click.

3.4 Correctness

You cannot perform a mathematically incorrect operation or derive a false result with Mathpert. It is not as widely known as it should be that most systems in use nowadays do not have this property. I shall therefore give some examples.

In Macsyma, we can set $a = 0$ and then divide both sides by a by using the command $\%/a$. This will result in $1 = 0$, because according to Macsyma, $a/a = 1$ and $0/a = 0$. This example can be carried out in Mathematica, too, but you must first tell it that dividing an equation means to divide both sides. The cause of the error is ignoring the side condition $a \neq 0$ on the rule $a/a = 1$.

Here's another example: if $F'(x) = f(x)$ then $\int_a^b f(x)\mathrm{d}x = F(b) - F(a)$. Apply this rule with $f(x) = 1/x$ and $F(x) = -1/x^2$ to get $\int_{-1}^{1}(1/x)\mathrm{d}x = 0$. The cause of the error is ignoring the side condition that f must be continuous on $[a, b]$. This is a standard homework problem in Calculus 1.

When these examples are presented, the suggestion is sometimes made that these

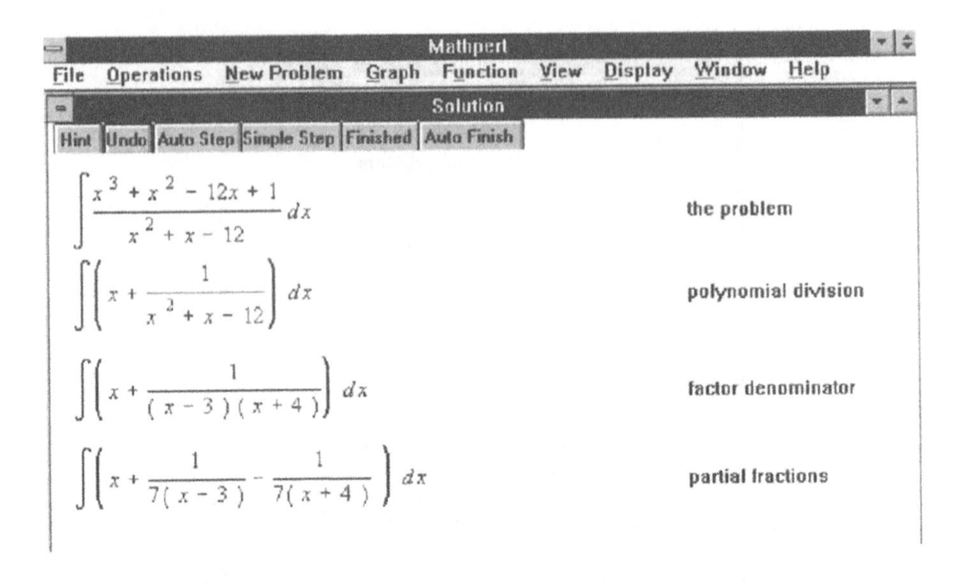

Fig. 3. A solution with some steps using powerful operations

are simply "little bugs" that can be easily removed. This is a misconception. The root cause of these problems is that mathematical operations generally have "side conditions" which are logical propositions. For example, the side condition on the operation $a/a = 1$ is $a \neq 0$. When we make a mathematical calculation, each line is actually dependent on a certain list of assumptions. Some steps, for example replacing $\sqrt{xy}$ by $\sqrt{x}\sqrt{y}$, add new assumptions to this list. (In this case, the new assumptions are $0 < x$ and $0 < y$.) When we do mathematics by hand, we usually do not write the assumptions down explicitly, but sometimes we make a note of additions to the list, and maintaining the list correctly is vital, as the above examples show. Moreover, it is not possible to restrict the complexity of these assumptions. Denominators must not be zero; but denominators can be complicated, so the zero-sets of functions get involved. These often depend on the domains of other functions, so we need to be able to calculate domains. Domains, for example of $\sqrt{u(x)}$, will depend on the sign of u in a neighborhood. This kind of question can in turn depend on whether functions are monotone increasing or decreasing; things get arbitrarily complicated. The side condition for evaluating definite integrals already involves continuity of the integrand on an interval. These complexities show that the correctness principle cannot be "added on" to a system that was not designed from the start to support the maintenance of a list of assumptions during a calculation.

The correctness principle applies to graphs as well as to symbolic solutions. When graphs have singularities, most commercial graphing software gets most of them wrong. Graphs are made by connecting a finite set of points. These points will probably not hit the singularities exactly, so the graph may not even go off-screen at a singularity. Moreover, there is often a vertical line where the singularity belongs. This arises because the software just connects one point to the next, and "knows" nothing about singularities.

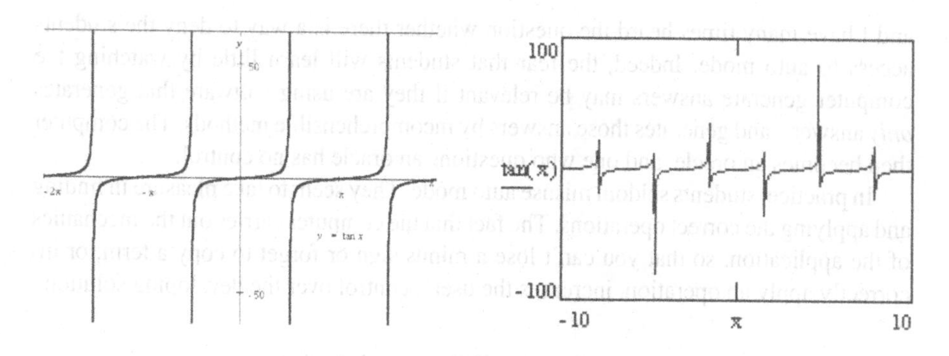

Fig. 4. Correct versus incorrect graphs of tan x

Mathpert avoids these pitfalls, by means that will be explained in Sect. 5. Figure 4 shows a graph of tan x made by Mathpert, and a graph made by purely numerical computation, as all other software I have seen does it. (Some graphing calculators trap tan x as a special case and get it right, but they don't get tan x^2 right, or even simple rational functions.)

Now that I've explained the difficulty of satisfying the correctness principle, let me also emphasize its importance. Although this paper deals with *educational* software, I find it disconcerting to think of flying on planes whose jet engines or flight-control systems were designed using mathematical software that can derive $1 = 0$ in two steps. If there are two-line contradictions, there may also be more subtle errors that do not stand out. Not every wrong answer will be obvious, even to an experienced engineer. However, let us restrict the discussion to education. Errors that are obvious to an engineer may not be obvious to a student. If the software draws graphs with vertical lines at singularities, for example, students may be easily confused.

3.5 User is in control

The user decides what steps should be taken, and the computer takes them. As in word-processing, the computer takes over the more "clerical" part. The principle that the user should direct the solution requires either a command-driven or mouse-driven interface. The principle of ease of use dictates a mouse-driven interface. When Mathpert operates this way, it is said to be in *menu mode*. The user selects an expression on the screen. This selection brings up a menu of operations. The user then selects an operation. These operations transform the current line and generate the next line. In this way a multi-line solution is built up. Earlier versions did not have the capability to select an expression from the screen, and a more extensive system of menus was used to select the operation; hence the name *menu mode*.

The principle that the computer can help you out when you are stuck requires there to be an automatic mode, or *auto mode*, in which the program can itself generate a solution. The user can switch between these two modes at will.

Each time I demonstrate Mathpert, and professors see the program generating step-by-step solutions to homework problems, some of them envision students simply having Mathpert do the homework for them while they are out watching *Beavis and Butthead*, and I have many times heard the question whether there is a way to deny the students access to auto mode. Indeed, the fear that students will learn little by watching the computer generate answers may be relevant if they are using software that generates *only* answers, and generates those answers by incomprehensible methods. The computer then becomes an oracle, and one who questions an oracle has no control.

In practice, students seldom misuse auto mode. They seem to take pleasure in finding and applying the correct operations. The fact that the computer carries out the mechanics of the application, so that you can't lose a minus sign or forget to copy a term, or incorrectly apply an operation, increases the user's control over the developing solution.

3.6 The computer can take over if the user is lost

A good symbolic-computation program for education should be able to solve (upon request) all the (symbolic or graphical) problems commonly given in algebra, trigonometry, and first-semester calculus. This is a very strong requirement, in view of the fact that what is required is not just the answer, but an economical, beautiful, and cognitively faithful step-by-step solution, tailored to the level of the student.

Even if we only wanted answers, however, it is important to note that the restriction to problems commonly given in courses is vital. There are technical results showing that it is impossible to write programs to solve *any given* problem of the types studied in these courses. For example, Richardson (1968) showed that it is impossible to write a program that will take as input a given identity $f(x) = g(x)$ in one real variable, and determine whether or not that identity is valid, where f and g are built up from exponentials, logs, and trigonometric functions using the arithmetic operations.[1] Nevertheless, we do implicitly teach methods, not just a collection of specific examples, and the requirement is simply that the program must embody all the methods we teach.

The ability of the system to generate an "ideal" solution can be put to good pedagogical use to assist a student who is having difficulty. Several such features are included in the design of Mathpert, but as these are not fundamental design principles, they will not be discussed in this section.

Problems often can be solved in more than one way, for which the jargon is "multiple solution paths". Of course the ability of the computer to generate a single "ideal" solution does not contradict its ability to support multiple solution paths in menu mode. But it actually offers good support for multiple solution paths in auto mode too. Although the system generates for each problem a unique "ideal" solution, it has to be able to do that from any point. The user could take one or several steps, and the computer must be able to finish up the problem from there. Usually the computer can be sent down a particular "solution path" in a few manually-directed steps, after which it will finish the solution along the desired lines.

Sometimes the necessity of solving a problem by trial and error is brought up in this connection. Mathpert can carry out such a search, as when searching for the factors of a

---

1 That is, there is no *decision algorithm* for this class of identities. Normally in trigonometry, the trigonometric functions are not nested, and the arguments of the trigonometric functions are linear. There is room for more careful work, narrowing the gap between the large classes of identities for which there is provably no decision algorithm, and the small classes for which we actually possess a decision algorithm. The "boundary" is at present largely unexplored.

quadratic polynomial. When this subject is first being learned, the attempted factorizations are shown to the student. There is a more extensive discussion of trial-and-error in Sect. 4.4.

3.7 Ease of use

The system must be easy to use. This point has been emphasized over and over by everyone who has considered the use of computers in mathematics education. Students are likely to be poor typists and may be afraid of computers. Many of them are also afraid of mathematics. It is essential that the students feel the computer is a help, not just another hurdle. Reports I have seen of many experiments with computer-assisted calculus instruction show that often the students do feel the computer adds to their difficulties, rather than helps. The day has not yet arrived when many students buy mathematical software that is not required, because they think it will help them learn.

Ease of use is often considered to be an "interface issue", but I don't believe that a well-designed program for education can really be separated into an "interface" and a "kernel" operating completely independently. Take the generation of step-by-step solutions, for example – is that an interface question? If you take the narrow interpretation, it is not. Only the means by which the user commands the steps to be taken, and the means by which the results are displayed or printed, are interface questions. But on this narrow construction of the term "interface", the question of ease of use cuts much deeper than the *mere interface*. Ease of use means that it is easy for the user to accomplish a given purpose. Whether the purpose can be accomplished easily, or at all, may well depend on the capabilities of the program as well as on the design of the menus and dialog boxes and which buttons can be clicked when.

A critical moment for "ease of use" comes when the user is "stuck", that is, does not know how to proceed. If the system has the capability to solve the problem, it can be viewed as an "interface issue": *how to use that internal capability to best advantage to help a student who is stuck*. Whether one has the program simply exhibit the next step, show the next menu choice, or give a hint in English (or French for that matter), is a question of interface design, in my view. I would not limit the term "interface" to mean where you have to click or what you have to type to cause these things to happen.

Nevertheless, questions of interface design even in the narrow sense are of some importance. I will consider some of them in Sect. 8.

3.8 Usable with standard curriculum

Even if the professor isn't using it, and knows nothing about it, it should be usable by the student. This criterion is perhaps more controversial than my other criteria. But I have always believed that serious curriculum change will be driven by student use of software, rather than the other way around. After a decade of well-funded initiatives to bring technology into the calculus classroom, most classes at most universities, and all classes at many leading universities, are still taught without software assistance, and with a substantially unchanged curriculum.

Answer-only programs such as Mathematica and Maple are hard to use with the standard curriculum in mathematics, which emphasizes step-by-step solutions. This has led some people to call for a new curriculum, in which students would not learn the traditional step-by-step solution methods. Methods of integration in particular have suffered a bad press. There is talk of emphasizing *concepts* instead of *techniques*. Of course, nobody is against teaching mathematical concepts – it's like motherhood and apple pie. But there is a serious question whether it makes sense to teach concepts without techniques.

Given the fact that answer-only software can't support the present curriculum, the faculty has been left with the choice: stick with the old curriculum (and no software), or take a big leap: changing both what is taught and how it is taught at the same time. Change will be easier if they can change gradually, adding software support to the existing curriculum, and then adjusting the curriculum later as required.

4 Cognitive fidelity and glass box in Mathpert: operations and pedagogy

The achievement of cognitive fidelity in Mathpert depends on the implementation of "operations" that correspond to identifiable "steps of computation" taken by a student in the course of solving a problem. These operations can then be invoked by name or formula from a menu (in various ways) or applied by the program itself when generating an exemplary solution. It is essential that the basic steps of the solution correspond to what the human perceives as "operations".

4.1 Design of the operations

The cognitive-fidelity principle requires a careful and detailed analysis of the subject matter, to determine the correct choice of operations. There are some 130 menus of up to sixteen operations each, so there are about a thousand different operations that Mathpert can perform (not counting graphical operations). I believe that these operations permit the solution of all (non-word) problems normally taught in algebra, trigonometry, and calculus. In many cases it was a nontrivial task to construct the appropriate set of operations, though in some cases it was not difficult for an experienced teacher.

4.2 Rewrite rules, matching, and pedagogy

A rewrite rule is a one-directional equation, such as $a(b+c) = ab + ac$. By saying it is one-directional, I mean that you can use it to rewrite $a(b+c)$ as $ab + ac$, but not the reverse. A rewrite rule is applied by using a process of *matching* to see that some complicated expression "has the form of" the left-hand side. For example, we can use the rule just mentioned to rewrite $x^4(x+1)$ as $x^5 + x^4$. Purists will note that I have used some other laws of algebra and arithmetic as well!

Although rewrite-rule technology at first sight appears attractive, many of the operations cannot be expressed in rewrite-rule form, because they take an arbitrary number of arguments and because other arguments can come in between. Consider "Collect powers", for example, which is related to the rewrite rule $x^n x^m = x^{n+m}$, but applies to $ax^2bx^3x^4cx$. The last x has to be collected even though it doesn't have an explicit exponent. The other x's have to be collected even though they are separated and there are more than two of them. Moreover, there are further difficulties with rewrite rules in the presence of associativity and commutativity.[2]

These observations have implications of interest to both programmers and educators. The implication for the programmer is simply that rewrite-rule technology isn't enough: each operation has to be implemented as a function in the programming language.

The implication for educators is more far-reaching, though less obvious. Many of the difficulties ("bugs") that I have observed in student solutions (or failures to generate solutions) can be attributed to failures in the student's matching algorithm (the "buggy theory of learning" was first promulgated by J. S. Brown and R. Burton, see, e.g., Burton 1982). Just to give one example: consider using the (reverse) distributive law $ab+ac = a(b+c)$ to rewrite $x^4(x+1)+x+1$ as $(x^4+1)(x+1)$. Even if the difficulty about grouping the first $x+1$ so as to match a to $x+1$ is overcome, there is still the problem that c is 1 here, so literal matching won't suffice. The matching procedure itself has to incorporate some other algebraic laws for this to work right. Another, unrelated difficulty apparent in this same example occurs the first time students have to match one of the letters in a pattern to a sum, instead of to just a monomial. Nearly every student fails the first time this is asked. Yet many textbooks I have examined fail to give an illustrative example. The first such problem occurs in the homework assignment, where it is guaranteed to be frustrating. Many a case of "math phobia" begins with examples like this, in which a homework problem is asked which in some sense "follows from" the principles in the text, but involves some new twist, which the computer scientist can describe precisely in terms of extra code required for the matching algorithm. [I am certainly not the first to observe that student errors are sometimes attributable to faulty matching. See, e.g., Nguyen-Xuan et al. (1993) and Nicaud et al. (1993).]

4.3 Order of operations

It is interesting, both for the computer scientist and the educator, to consider the problem of which rule to apply when. First of all, one should observe that many rules are inverses: for example $\log ab = \log a + \log b$ can be used in either direction, and indeed *must* sometimes be used in one direction, and sometimes in the other direction. Sometimes we must factor, sometimes we must multiply out.

The simplest control mechanism would be context-free, in the sense that the decision as to which operation to perform could be made *locally*. For example, the decision

2 Readers with training in the relevant branches of logic or computer science will be aware that there is a vast literature on rewriting techniques. One of the general aims of this research is to provide improved matching (technically: unification) algorithms which will incorporate algebraic laws into the matching step, so that for example $ab+c$ should match $c+ba$. Although there are beautiful algorithms and theorems about those algorithms, it should be noted that the best of the crop will still not handle the rewriting of $x^4(x+1)$ as x^5+x^4. Readers wishing to explore these algorithms will find an initial discussion and further references in Bundy (1983).

This body of theory has not found application in Mathpert, and not because the author was ignorant of the theory. Nevertheless Mathpert successfully handles matching. For example, it is able to apply the rule $\cos^2 x + \sin^2 x = 1$ in the context $2 + 3\sin^2 5x - 5xy + 3\cos^2 5x$. But if the second $5x$ is made $x5$, Mathpert will miss the match until the operation "Order factors" is applied. Note, however, that associative-commutative unification will miss the match in the first example until the 3 is factored out.

whether to put a sum of fractions over a common denominator would have to be made without regard for the context in which the sum of fractions occurs. In practice, there are two separate factors influencing such decisions that are not directly available in this method. First, it matters what kind of problem we are solving. For example, if the problem type (the goal of the problem) is common denominators, obviously we want to use common denominators regardless. If the problem type is partial fractions, equally obviously it would not be a good idea. Second, the context of the term being simplified (in the sense of the current line in which it is embedded) matters: for example, we never want to use common denominators inside an unevaluated derivative. It looks silly to use common denominators inside a derivative, like this:

$$\frac{\mathrm{d}}{\mathrm{d}x}\left(\frac{1}{x}+\frac{1}{x^2}\right),$$

$$\frac{\mathrm{d}}{\mathrm{d}x}\left(\frac{x+1}{x^2}\right)$$

followed by a long computation using the quotient rule; the "correct" solution differentiates term by term, without using common denominators. There are many instances of such contextual dependencies.

Adjusting Mathpert's problem-solving algorithm to take these contextual demands into account has accounted for much of the work on the algorithm. None of this would be necessary in an answer-only program. On the other hand, a great deal of fine-tuning is required to cause the algorithm to use the right operations in the right order to produce the right step-by-step solutions. During the process of developing Mathpert, I observed many loops and infinite regresses generated by meta-rules that conflict in a given situation. Each time I had to modify the meta-rules which looked good before, adding or changing the conditions under which certain mathematical operations should be applied.

Loops

Another complication, in addition to contextual dependencies, is the possibility that different operations taken in combination can cause loops. For example, you can factor and then expand a polynomial; so auto mode must never activate both operations, or a loop will result. There is nothing like a formal proof that the current system is always terminating; such a proof would be extremely complicated and probably could only be carried out with mechanical assistance. Indeed, it may well not even be the case – there may well be loops that haven't been observed. Certainly I have observed many unanticipated loops in the process of development and testing.

In order to guard against unanticipated loops in auto mode, Mathpert incorporates automatic loop detection: it will never go into an infinite regress, but will detect that it is about to enter a loop. Early versions would then deliver a rueful message: *I seem to be stuck in a loop. You'd better take over.* The present version does something more like what a person would do: it just doesn't use the operation that would cause a repeated line, but looks for something else to do.

The avoidance of loops is particularly critical in equation solving. The difficulty is that various algebraic laws have to be applied in one direction for some equations, and in the reverse direction for other equations, so that at first sight it seems impossible to avoid loops and still solve all equations. Here I have had some help from A. Bundy's theory of equation-solving (Bundy 1983). His general theory had to be integrated with special-purpose solvers for linear equations and non-linear polynomials, but his ideas of "attraction" and "collection" guided the way auto mode selects operations according to the form of the equation.

4.4 Pedagogical implications

Once algebra, trigonometry, and calculus have been codified into a set of about a thousand rules, can we say, "If you know the rules, you know the mathematics"? This is true to the extent that by those rules all the symbolic problems in the books can be solved. But it overlooks the fact that you must not only know the rules, you must know which rule to apply when. This is really much the more difficult thing to know. Mathpert has several thousand lines of what amount to *rules for using the rules*. These rules are internal to the program, rather than visible to the user, and govern the machinery of the automode solution-generator. The process of developing and fine-tuning these meta-rules took much more time than the implementation of individual operations for computation, which was comparatively simple.

Yet, when we look at textbooks and course outlines, most of what students are taught is how to execute the rules one or two at a time. Each section presents another few rules, and exercises which are solved by using those rules. The student studies algebra this way, and then learns trigonometry a few rules at a time. When the student comes to trigonometric identities, for the first time a variety of different rules is required. A detailed analysis by A. Lauringson revealed 59 operations required to solve a certain set of identities, while corresponding sets on other topics required at most 30 operations. Probably the notorious difficulty of this subject is mainly attributable to this feature. There is little or no discussion in textbooks about strategy for trigonometric identities. For example, there are three rules for expanding $\cos 2x$, and success is often dependent on choosing the right one, but textbooks do not even discuss the issue.

Solving equations is another area in which strategy is crucial, but seldom explicitly discussed. Here is a place where the contributions of computer science should be allowed to "trickle down" to textbooks. Bundy's ideas of attraction and collection are simple and straightforward and can easily be explained to algebra students. I am resisting the temptation to prove this point by explaining them here and now; but see Bundy (1983). These ideas bring order to the otherwise mysterious business of solving equations, reducing what appears to be an art to a science. Don't write another algebra textbook without reading Bundy.

Then the student goes into calculus, where he or she is required to "simplify" expressions in order to solve calculus problems. One semester I kept statistics on examination errors: Eighty percent of examination errors in first-semester calculus were actually errors in algebra or trigonometry that theoretically were prerequisite to the course. These errors were sometimes due to failures in the matching algorithm, sometimes to incorrect memory of a rule pattern, but often simply to not recognizing which rules would do the required job. *That* was never taught; the "born mathematicians" pick it up somehow, by osmosis.

The above remarks should be carefully considered not only by designers of such software, but especially by textbook authors, course designers, and teachers. No wonder students are confused: the control apparatus is more complex than the rules being controlled, but is never explicitly discussed! Textbooks should contain explicit discussions of these matters, as well as carefully chosen examples and exercises to illustrate the correct choice of operations according to the goal and the context. Most important, they should contain many more sections and examples in which a mixed bag of different rules need to be applied.

One interesting question is this: how much does mathematics depend on trial-and-error search? My experience in developing the auto mode algorithm of Mathpert leads to the conclusion that search is used in elementary mathematics in only a few places: searching for the factors of a quadratic polynomial, searching for a linear factor of a higher-degree polynomial, and searching for a good grouping of the factors in factor-by-grouping. In calculus we might add, searching for a substitution in integration by substitution, and searching for a choice of parts in integration by parts. All other problems, including the choice of which rule to use on $\cos 2x$, can be solved without searching. [Nicaud et al. (1993) describe a system that helps students learn to solve factorization problems by teaching them to search for a solution, backtracking at failure. The examples given in the paper are in the topic known in Algebra I as "factor by grouping".]

One of the referees raised the possibility that Mathpert could teach or give access to its internal meta-rules. This suggestion has two answers. The short answer is that Mathpert is not designed to be explicitly tutorial (see Sect. 1). It would almost certainly be a good idea to give more instruction in problem-solving strategy, possibly aided by more explicit "rules of thumb" suggested by the internal Mathpert meta-rules. However, it is not within the scope of the statement of purpose of Mathpert given in Sect. 1. The long answer is more interesting, and bears on deep issues in artificial intelligence and cognitive science. Namely: it is not at all clear that human mathematicians really operate in this way, using a collection of meta-rules. Maybe there is planning and goal-seeking at a deeper level; maybe there is "understanding". Chess programs operate by rules and calculations, but perhaps grand masters do not. Even if this is true, students of chess do study rules. This issue has caused more arguments than any other in the philosophical foundations of artificial intelligence.

5 Customizing the solution in Mathpert

In Sect. 2 we have given an example to illustrate the fact that Mathpert can generate solutions with many small steps, or solutions with few but powerful steps. This section will discuss how this is achieved.

First of all, when the user is controlling the steps of the solution, the user can choose the operation, and so it will suffice if there is a choice of the powerful or the weak operation. When the user has selected a sum containing a fraction or fractions, the menu of available operations will provide more than one choice. For the beginning algebra student (that is, if the problem was entered under an elementary topic), this list will include "Find common denominator". This will simply multiply both fractions by an appropriate factor c/c, as in the first step in Fig. 2. If the user thoroughly understands common denominators and basic algebra, for example a second-semester calculus student, the operation "Common denominator and simplify numerator" will do everything:

for example,

$$\frac{1}{x^3+1} + \frac{1}{x^2-1}$$

will be transformed to

$$\frac{x(x^2+1)}{(x-1)(x+1)(x^2-x+1)}.$$

A student still wanting to see more detail can first explicitly factor the denominator, and then use "Common denominator", and finally "Simplify", to get a three-step solution.

Customized solutions like this are easy for Mathpert to provide in menu mode, because the symbolic-computation engine was developed in accordance with the design principle of cognitive fidelity. Another project whose progress I have followed, was originally based on the idea of putting a step-by-step interface on Reduce. After the implementation had progressed for some time, it was noticed that the steps were often too large and confusing. Note that this is a completely different class of difficulties from the logical problems discussed in connection with the correctness principle. Once one starts down the road of implementing simpler operations in terms of more complex ones, using a programmable computer algebra system such as Reduce or Mathematica, what guarantee is there that the process will be simpler than implementing an educational computer algebra system from scratch? And even if it turned out to be, one would still have the logical difficulties.

Mathpert is designed not only to permit the user to produce different solutions to the same problem, but also to do so itself when running in auto mode. If a common denominator is required to solve an integral, it should be done in one step for a calculus student, but for a beginning algebra student, auto mode should produce the long solution to $1/x + 1/y$. How is this to be accomplished?

In order to support the generation of different solutions to the same problem, Mathpert maintains an internal model of its user. This model has been discussed in some detail (Beeson 1990); for this paper it will suffice to know that the model consists essentially in labelling each of the approximately one thousand operations as either *unknown*, *learning*, *known*, or *well known*. Initializing the model consists in assigning these labels.

Experience since 1990 has taught an important lesson. The original intention, inspired by the "buggy model of learning" (Burton 1982), was that the user model should model a particular individual. This model was to be initialized by a diagnostic test program, and the model was even made user-editable in early versions of Mathpert. However, experience soon showed that it is easy to get an "out-of-balance" user model, which will cause strange-looking solutions, in which there may be several tiny steps and suddenly a mysterious large step. Indeed, reflection shows that we don't actually want to model the *real* student, but rather some fictional *ideal* student at the appropriate level for the subject, so that the auto-mode solutions are ideal solutions, rather than ones which might be generated by our (buggy) real student. [A team working on the APLUSIX project has developed a program which diagnoses a student's knowledge in a limited sub-domain of algebra, based on some twenty or thirty transformation rules in polynomial arithmetic (Nguyen-Xuan et al. 1993, Nicaud 1994). This work represents the state of the art in diagnosing and modelling individual students in the subject of mathematics.]

The end result of these considerations is that the user model has become invisible to the user. When the user first enters Mathpert, he or she chooses from a menu what the "topic" of the session will be. These topic choices are intended to encompass about one day's lesson each (corresponding to one section of a textbook). The user model is automatically initialized based on the choice of topic. This user model will produce solutions at the level of detail generally appropriate for a class on that topic; the solutions will not use operations that are unknown (unless the problem can't be done without them). Thus L'Hôpital's rule will not be used (in auto mode) on $\lim_{x\to 0}(\sin x)/x^2$ when L'Hôpital's rule won't be taught until next week; but next week, when the topic is L'Hôpital's rule, it will be used.

In short: it suffices to customize the solution to *the level* of the individual student. It doesn't really need to be customized to the *individual* student.

6 The correctness principle in Mathpert

Necessity of a theorem-prover

Having established the difficulty and importance of the correctness principle, let me remark on what it took to build a system that satisfies it. Mathpert incorporates a non-trivial theorem-prover (about 6000 lines of C in the current implementation, about ten percent of the total code). This prover is a piece of work that can stand on its own as a contribution to the branch of computer science known as automated deduction. There are few other programs combining some symbolic-computation capabilities with logical-deduction capabilities; we may mention Wu's (1986) work on geometry and Analytica (Clarke and Zhao 1992) in this connection. Descriptions of the Mathpert prover can be found in Beeson (1989, 1992); and a discussion of some later additions is in Beeson (1995). (I have recently written another program called Weierstrass, based on the Mathpert prover, which has automatically generated an epsilon-delta proof of the continuity of x^3; previous programs could do this only for linear functions.)

Logic is in the background

Although a correct handling of logical matters is essential for correctness, we rarely ask students to consider logical matters explicitly. The cognitive-fidelity principle then requires that the logical work performed by the software remain mostly invisible. The glass-box principle, however, requires that it be visible on demand. Each (visible) line of the solution has associated with it a list of assumptions on which it depends. For example, if you enter $1/x + 1/y$ as the problem, this will depend on the (implicit) assumptions $x \neq 0$ and $y \neq 0$. Mathpert makes these assumptions explicit, but does not display them unless you request it.

One of the referees questioned this point of the design, saying that one of the main points of undergraduate mathematical education is training in logical thinking, and that therefore attention should be focussed on the logical aspects of the steps. Personally, I think this is a valid point deserving of consideration by teachers and textbook authors. Indeed the "View assumptions" option in Mathpert can certainly be used to expose the logical underpinnings, so the teacher could simply instruct the student to make use of it.

Moreover, on those occasions when Mathpert refuses to execute an operation because its side conditions cannot be satisfied, the user's attention will be forcibly drawn to the logical aspect of the situation. But the design principle of supporting the existing curriculum took precedence in this case (as well as some others) over the desire to "improve" the existing curriculum according to my own lights.

Correctness of graphs

As discussed in Sect. 2, a purely numerical approach to graphing leads to incorrect graphs when the function being graphed has jumps or singularities. Graphs are made by connecting a finite set of points. If these points are chosen simply by partitioning the x-axis, the points will probably not hit the singularities exactly, so the graph may not even go off-screen at a singularity. Moreover, there is often a vertical line where the singularity belongs. This arises because the software just connects one point to the next, and "knows" nothing about singularities. Similarly, jumps will be drawn as short nearly-vertical lines. There is another class of function that is often incorrectly graphed: those with many maxima and minima close together. Two common examples would be $\sin(40x)$ and $\sin(1/x)$. What happens to a purely numerical grapher is that when the peaks are sharp, a maximum will occur between two plotted points, and the line segment connecting the points will simply cut across, going nowhere near the true maximum. Even an "adaptive step size" algorithm such as is used by Mathematica will not help, because the numerical clues that the step-sizer looks for occur only near the peak, and they are skipped too! Matters are made worse by the fact that these errors sometimes produce highly symmetric patterns, e.g., in the case of $\sin ax$ for a between 30 and 50. It's an interesting puzzle to explain the patterns, but they bear no relation to the true graph, and may well confuse a student.

Mathpert solves these difficulties by making use of its symbolic-computation engine both while preparing to graph and during the actual graphing. Let us consider the problem of sharp maxima first. First of all, before graphing Mathpert computes the symbolic derivative of the function to be graphed, and then, as it graphs, it numerically evaluates the derivative as well as the function. If the derivative changes sign from one plotted point to the next, Mathpert uses a numerical equation-solving method to find a zero of the derivative between the two points, and makes sure to plot a point at that x-coordinate as well as the original two. In this way it ensures hitting the maximum. Of course, if the graph has maxima closer together than one pixel, some inaccuracies are going to result anyway. After all, there is no hope of drawing a perfect graph of $\sin(1/x)$ on a screen with a finite number of pixels! But what we can achieve is a clear picture of the envelope of the graph.

Turning to the problems of singularities and jumps: Mathpert appeals to the symbolic code used by the theorem-prover (not the operations accessible to the user) to calculate the singularities (if any) before making the graph. It then takes care to draw the graph correctly at the singularities, now that it "knows" where they are. For example, it calculates that the singularities of $\tan x$ are at $(2n+1)\pi/2$, and then, when graphing $\tan x$ over a specific interval $a \leq x \leq b$, it calculates the relevant values of n and draws the graph correctly.

In difficult cases the program may fail to find a formula for the singularities. Even if it does find a formula for the singularities, if that formula involves an integer parameter as in the above example, it may be too difficult to find the set of relevant values of n. That set may not even be finite. For example, $\tan(1/x)$ is too hard. But, Mathpert at least graphs all rational functions correctly, and I hope it graphs all functions that would arise in a calculus course correctly.

It is easy to adapt the result of Richardson (1968) to show that the problem of computing the singularities of a given elementary function is not recursively solvable. No matter how one improves the algorithm, there will always be functions whose singularities are not calculated. However, in such a case the user will get a warning that Mathpert could not calculate the singularities, and the graph may be incorrect.

Another approach to these problems, "honest plotting", is reported on by Avitzur in this volume and also by Avitzur et al. (1995). The term "honest plotting" was introduced years ago by Fateman (1992) and refers to the use of interval arithmetic in calculating the graph values. That is, all arithmetic is done carrying along upper and lower bounds. In places where the possible error is significant, the graph line gets thicker. In bad examples, such as $\sin(x)/x$ near the origin, it becomes a region instead of a line, but at least the correct graph is known to lie in the shaded region.

The correctness principle and pedagogy

I found in the writings of Maria Montessori (1967, chap. 24), the Italian educator, a principle that is highly relevant. Montessori demanded that educational materials satisfy the principle she called "control of error". This means that the materials should be "self-correcting": the materials themselves must inform the student, without the intercession of the teacher, when the activity has been completed successfully. Montessori had in mind physical materials for preschool and elementary-school children. For example, suppose the task is to sort ten wooden dowels by length. Control of error can be provided by making the rods of different diameters, and requiring the sorting to be done by inserting the rods in holes. You can't put a rod in a too-small hole. You can, of course, put a rod in a too-large hole, but if you do, you'll have some rods left over that won't fit anywhere. There is only one way to fit all the rods in a hole, and that's the correct way.

Mathpert provides control of error by means of the correctness principle. You simply can't take a mathematically incorrect step. You can, of course, take a mathematically correct but irrelevant step; you can factor a polynomial and then multiply out the factors again. You can go on repeating those steps all day long if you like, but you can't make a mistake when multiplying out the factors. When the problem is finished, Mathpert will tell you, *That's the answer*.

7 Using the computer's power when the user is stuck

One of the design principles in Sect. 2 required that the computer should be able to take over on request (presumably, that means when the user is lost). There are, however, a number of intermediate and pedagogically interesting states the user can be in besides "proceeding competently" and "lost". There is, for example, the state "now what?", in which the user has been proceeding competently, and is generally competent with this topic, but does not see immediately what to do next. Even the state "lost" can be divided into "lost and given up" and "lost, but hoping to recover".

How can we best use the power of the computer to solve the problems, when the user is in these various states of confusion? This section will explain how the current version of Mathpert uses its auto-mode capabilities in different situations.

First of all, the user who is "lost and given up" can simply click the "Auto Finish" button to let the computer finish the solution. Presumably the lost student then studies the solution, in order to do better next time.

The user who is "lost, but hoping to recover" can click "Auto Step". This causes Mathpert to generate just one (more) step of the solution. Repeated clicking on "Auto Step" will thus duplicate, one step at a time, the effect of "Auto Finish". But perhaps the student can realize his hope of recovery after seeing one or two steps, and continue the solution himself.

Hint generation

A more interesting case is the user in the state "Now what?", who does not quite feel lost, but can't quite figure out what to do either. This user could of course click "Auto Step", but may be reluctant to "admit defeat" by so doing. Mathpert provides another alternative: the "Hint" button. When the user presses "Hint", Mathpert internally generates one more solution step, but does not show it. Instead, having determined what operation it would use, it looks up an appropriate natural-language hint in a pre-stored table of hints. These hints are designed to sound like what a teacher would say in the situation. They are conversationally phrased, containing formulas only when necessary, but they are designed to enable the student to extract the menu choice which should be made to invoke the suggested operation, or at least the term to select to proceed by using the Term Selection interface.

Error analysis

Although you can't carry out an incorrect step in Mathpert, you can try to do so, by making an inappropriate menu choice. Of course, the operation may be applicable even though in some sense inappropriate, in which case Mathpert, in accordance with its principles, will let you take that step. But you may have chosen an inapplicable operation, in which case Mathpert must refuse your request. It is desirable that in this situation the most informative error message possible should be supplied. Of course, some very common errors can be trapped individually, and appropriate messages can be stored in the program. However, with over a thousand operations, there are a million possible pairs of the form (*correct operation*, *chosen operation*). So some attempt at dynamic error-message generation is imperative. For example, some errors are simply due to omitting a necessary preliminary step. Mathpert will catch many of these errors. The method is this: when an operation is inapplicable, Mathpert internally takes four steps of the auto-mode solution starting from the current line. If the user's operation is used in one of these four steps, it is a good bet that the user is "on the right track", and a helpful message can be generated informing the user what preparatory step has to be done first. Of course, errors that are "not in the ballpark" will still have to generate some uninformative message such as *Sorry, that operation can't be applied here*.

8 Traditional interface issues: ease of use

In this paper I have focussed attention on the way in which the interface and "kernel" or computational engine of a good symbolic-computation program for education are *necessarily* interrelated. Nevertheless, there are a number of issues, mostly connected with my design principle, ease of use, that are traditional "interface issues".

8.1 Source of problems

Although Mathpert can solve problems given by the user (instead of only pre-stored problems), I found when our student laboratory opened that nobody actually wants to type problems in. Both students and professors want to find the problems already in a problem file ready to call up. Moreover, if using the program in connection with a class, they want to find the exact problems that have been assigned in class, not the problems chosen by the author of Mathpert. This is important because it permits professors to make the use of Mathpert optional. This problem has been solved, in a way that gets appropriate problems to the student without even requiring file names to be known or typed, while still allowing full flexibility to either student or professor to customize the problem files as desired.

When a student installs her new copy of Mathpert, she will choose her textbook from among the supported textbooks, and the problem files for that textbook will be included in the installation, in addition to the problem files developed and distributed with Mathpert.[3] Nevertheless, the program has to support entering an entirely new problem of your own choice. At present this means typing things like `x^2+y^2` when you mean x^2+y^2. There is thus a "one-dimensional notation" for typing in mathematics, different from the "two-dimensional notation" used in books and for on-screen display. Some other programs offer "equation editors": click-and-drag interfaces, using palettes of symbols, that allow users to build up formulas directly in two-dimensional notation. The construction and desirability of such interfaces is today a widely discussed issue, to which I have little to contribute. What little I have is simply this observation: while an engineer is likely to bring her own problem, a student is likely to want to solve the homework problems. If they are already on disk, nobody will have to type (or click and drag) to get the problems entered. Figure 5 shows what it looks like to see the problems stored on disk. You can press "Next" or "Previous" to peruse the available problems.

8.2 Selection of operations

There are over a thousand operations in Mathpert. Even with a menu-driven interface, if all thousand operations were always available, it would be confusing to try to find the one you need. There is therefore a serious "interface problem" in enabling the user to find an appropriate operation and cause it to be used. It seems useful to distinguish here between the user who knows what she wants to do, and the user who does not. Let us consider the user who knows what she wants to do, by which I mean that she knows which operation she wants to apply. This is not the same as knowing the name by which

3 As of March 1995, only two textbooks are supported, and we are in the process of obtaining permission from publishers to support more.

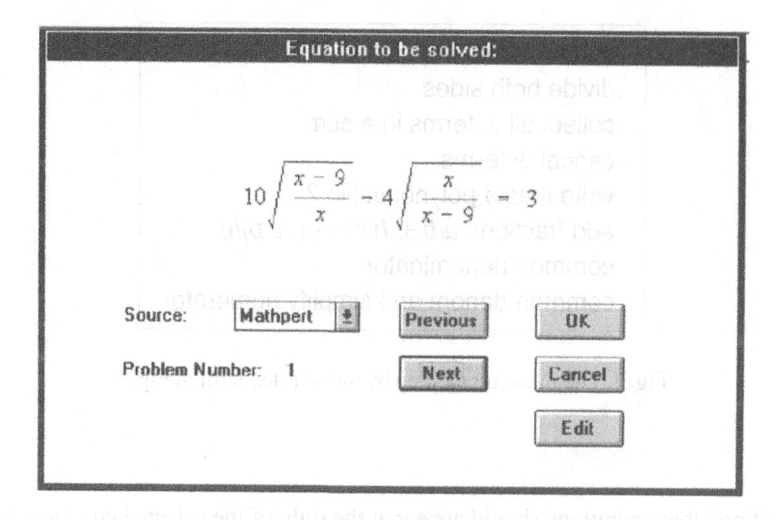

Fig. 5. Getting a problem without typing

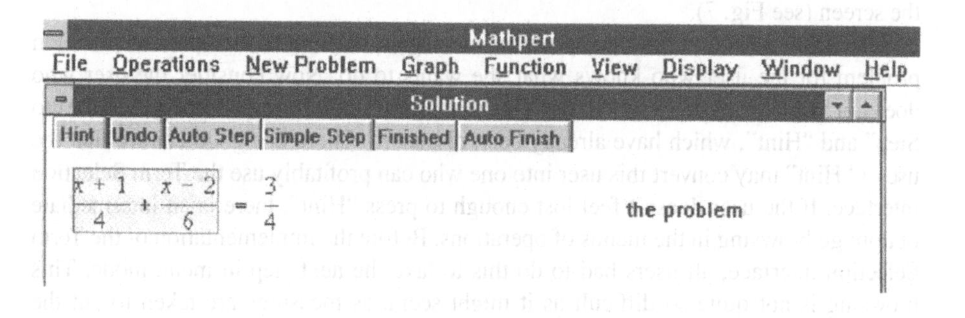

Fig. 6. Selecting a term to work on

this operation is known on a menu; what is meant is that she could easily write out the next step with pencil and paper, unless the computations are too intricate to do by hand.

In particular such a user will know which sub-expression she wants to work on. Mathpert permits her to *select* this expression by using the mouse to enclose it in a rectangle. When she makes this selection, the enclosing rectangle gets a colored background, just as when you select text in a word processor, and a short menu of operations comes up, sometimes containing only one operation, but sometimes containing three or even ten operations. The listed operations include everything you could do *to* the selected term, and they include also everything you could do *with* the selected term, such as transfer it to the other side of the equation. This Term Selection interface allows the user who knows what she wants to do to get the next line on the screen as quickly as possible, with a minimum of menu-searching. Figure 6 shows how the selected term appears on the screen. The selected term is highlighted (the default color is light yellow, but all colors can be changed by the user, and of course the illustration here is monochrome). The

subtract from both sides
divide both sides
collect all $\pm$ terms in a sum
cancel $\pm$ terms
write it as a polynomial in ?
add fractions $a/b \pm b/c = (a \pm b)/c$
common denominator
common denom and simplify numerator

Fig. 7. The menu brought up by the selected term in Fig. 6

menu of possible operations should appear at the right of the selected equation, but the image-capturing software used to make the illustrations proved incapable of handling a "floating menu". Therefore the menu has just been typeset, rather than captured from the screen (see Fig. 7).

The Term Selection interface provides an elegant solution to the operator selection problem for the user who knows what she wants to do. Now consider the user who does not know what she wants to do. This user has several options, among them "Auto Step" and "Hint", which have already been discussed. There is a good chance that the use of "Hint" may convert this user into one who can profitably use the Term Selection interface. If the user doesn't feel lost enough to press "Hint", there is an intermediate option: go browsing in the menus of operations. Before the implementation of the Term Selection interface, all users had to do this to take the next step in menu mode. This browsing is not quite so difficult as it might seem, as measures are taken to cut the number of menus of operations that can be browsed to a minimum. Luckily, in any given mathematical situation, most of the operations are irrelevant. Normally, all the operations you might need will fit on fewer than 20 menus of at most 16 operations each, so you need only one menu with entries like "Fractions", "Square Roots", and "Integration by Substitution"; each of those has an associated pop-up menu listing the operations themselves. Of course, this menu of 20 menus must be dynamically computed from the problem type and the current line of the computation – it can change after each new line. For example, when the last integral sign is gone, no more integration menus will appear.

The Term Selection interface is quite new, and all laboratory experience with Mathpert was based on the menu-browsing interface. Two facts emerged: (1) it is feasible to use this interface, indeed students quickly develop the skill of finding the operation they want; (2) nevertheless, nobody really likes doing so, although they do like seeing the solution develop as a result. In other words, the menu-browsing is a psychological "cost" to be weighed against the "benefit" of results obtained. The Term Selection interface, on the other hand, is fun to use.

Incidentally, multiple selections are allowed; for example one can move several different terms to the other side, or one can apply the same operation to two different subexpressions at the same step.

8.3 Arguments of operations

Some operations require an argument, that is, an additional term to be supplied by the user. For example, the operation "Add to both sides", used in solving equations. If the user chooses this operation in the old menu-browsing interface, a pop-up window will appear for data entry, and she would be prompted, *Add what to both sides?* After she enters a term (and it is checked and accepted as a suitable response for that particular operation), then the pop-up window will disappear and she can continue. In auto mode, if an operation requires an argument it must be automatically selected.

The Term Selection interface improves this markedly: to multiply an equation by something, just select that something in the visible equation, and then choose "Multiply both sides" from the resulting short menu. This works fine as long as the desired argument is visible somewhere to be selected. In some cases multiple selection can be used too, as in the case of multiplying an equation by 12 because it has denominators of 3 and 4 in different places: just select both the 3 and the 4. But if for some reason you want to multiply the equation by 27 when no 27 occurs, you will have to do that the old way, by selecting the whole equation, then choosing "Multiply both sides", and then typing in 27 when prompted.

The necessity of entering arguments for operations means that the problems of one-dimensional versus two-dimensional notation can't be swept under the carpet entirely by providing problems on-disk. Fortunately, the arguments to operations are typically quite short and simple. Moreover, they often (perhaps almost always) are expressions currently visible on the screen, and so the Term Selection interface can be used. The remaining cases when somebody has to type in an argument should be very rare. I am unable to give an example.

8.4 Justifications

The operations must return not only the result of the calculation, but also a justification for display as the "reason" for the step. An interesting and subtle interface question arises here. On the one hand, these "justifications" must look right when printed in a solution, as a reason for a step. On the other hand, they should also correspond to what is written on the menu to invoke the corresponding operation, so that if one has a Mathpert solution in hand, one could in principle duplicate it. For example, if the student takes one step in auto mode, it should be apparent how that step could have been taken in menu mode. These criteria were not always easy to balance. For example, "Factor out highest power" might occur on a menu, and one might want the justification to be $ab + ac = a(b + c)$. But this justification won't lead the user to "Factor out highest power". Careful attention to both desiderata while making up the justifications was necessary, and perhaps the best balance has not always been struck. With the introduction of the Term Selection interface, this has become a less important issue, as the necessity to browse the menus has diminished considerably.

8.5 Interface to the grapher

Mathpert can make every kind of two-dimensional graph. It can make ordinary graphs of a function $y = f(x)$; it can graph two functions at once (on the same axes or on different axes); it can make polar or parametric plots. It can graph not only functions but arbitrary relations $f(x, y) = 0$. It can also show you the complex roots of a polynomial, or solve an ordinary differential equation.

It is one thing to make a grapher and another thing to make a grapher that is usable for classroom demonstrations and independent use by students. You must be able to get the graph you want quickly. You must be able to show the effect of changing a parameter *quickly*, without having to go through an elaborate data-entry system first. For example, we want to draw a series of graphs of $y = x^3 - ax$ for $a = 1, 2, 3, 4, \ldots$, and then $a = 1, 0, -1, -2, \ldots$

You must be able to "zoom up" to a larger portion of the graph or "zoom down" (microscope style) at will, without data entry and waiting. You should be able to change every detail of the graph's appearance (line width, colors, etc.), yet not be distracted by screenfuls of such information when you don't want to change these things.

Some users reported to me that they find it important to be able to put a grid over the graph and remove it at will. People want to be able to make the ticks on the x-axis occur at multiples of π when they are graphing trigonometric functions.

The ability to change parameters and axis ranges quickly is vital to education. For example, given a grapher that can plot the complex roots of a polynomial, if you can change parameters rapidly you can do the following: show how the roots of $x^2 - bx + 1$ move when b is changed, at a speed fast enough that the movement appears to be animated. (They move around the unit circle, through a bifurcation where they meet at $x = 1$, then spread out in opposite directions on the x-axis.) One can also make n a parameter and see how the roots of $x^n - 1$ depend on n.

Small details of an interface often make a difference to the "look and feel" of software. An example that comes to mind is the question of putting titles on graphs. People want the title of the graph to use normal (two-dimensional) mathematical notation. That means it won't fit into a one-line title bar and should be placed on the graph itself. But then, the user must be able to move the title (with the mouse) to the location where it "looks best". This will vary from graph to graph. Moreover, if two graphs are being drawn on the same axes, the two titles need to be independently movable, so they can be placed near the corresponding curves.

Given the above requirements, what is the correct interface design? One wants to provide tools to accomplish the above tasks, in such a way that users will see and use the tools, but without cluttering up the screen, and adhering to the design principle of ease of use. Since there are so many things you can do to change your graph, it's not easy to provide a complete set of tools for accomplishing all those things *easily* without overwhelming the user with too many confusing icons, buttons, and menus. Mathpert tries to provide a small number of the most commonly-used tools in a visible form as a Graph ToolBar, occupying a narrow vertical strip to the left of each graph. The rest of the tools are accessed via menus. Using the ToolBar, you can zoom in or out and change parameter values with a single click.

One interesting issue in the design of a grapher interface is whether the graph should be instantly updated when a change is specified, or whether the user should be able to specify several changes before redrawing. The increasing speed of computers that run Mathpert has weighed in favor of instant redraw, for example when a zoom button or parameter-increment button has been pressed. Avitzur's (1995) Graphing Calculator sets a high standard for a grapher interface, and the Power Macintosh on which it runs certainly has enough speed to make instant redraw attractive. But there are a few circumstances where instant redraw is awkward, namely when the redraw is not actually instant, but takes too long. For example, looking at the partial sums of a series with the number of terms as a parameter, you will not want to change the number of terms from 5 to 50 by increments of 1 if you have to wait for each graph along the way to be drawn. (Such a formula can't be entered in the Graphing Calculator.) Another example is a contour plot. In most situations, however, instant redraw is preferable, so the cases mentioned have to be handled as well as possible by other features of interface design, such as easy access to dialog boxes for changing the parameter value by large amounts, and a button to interrupt a slow drawing.

9 Interfaces and pedagogy

The development of software for symbolic computation has already caused substantial debate about the mathematics curriculum and pedagogy. The development of education-specific software for computations and graphs will have further effects.

Here is a single example of what can be done. When teaching Calculus II by using Mathpert, one day I introduced exponential functions $y = a^x$ graphically, letting them change the parameter a. I asked each student to find the value of a which gives the graph slope 1 as it crosses the y-axis. They had to read the numerical value of the slope (which is easy with Mathpert), and adjust the value of a accordingly. In this way they each soon arrived at value the of the exponential accurate to several decimal places. Then we used the graphical ordinary-differential equations solver to look at the solutions of $y' = y$, and saw their similar shape. We then calculated the derivative of e^x symbolically, using the definition of the derivative as a limit. Finally they were given a file of homework problems on differentiation of functions involving exponentials.

I promised at the beginning of the paper to return to the issue of software that would not only replace pencil and paper, but also teachers and books. One preliminary step in this direction is afforded by the Microsoft Windows technology known as OLE, Object Linking and Embedding (Brockschmidt 1994), and by the competing Apple technology OpenDoc. This means that documents created by one software application (such as a spreadsheet or Mathpert) can be embedded in a document created by another application (such as a word processor). In practice it works like this: You write a page of a textbook in your favorite word processor explaining the exponential function $y = a^x$. You make a graph of it in Mathpert and paste that into your document with tools that today come with Windows. You include instructions for the student to click on the graph and find the value of a that makes the slope be 1 where it crosses the y-axis. The student reading your document clicks on the graph, and Mathpert starts up on that document. When the student has completed the exercise, she exits Mathpert and is back in your textbook, ready for the next part of the exercise, as described above. OLE technology means that anyone who can write textbooks can also write teaching materials that incorporate symbolic-computation software. Publishers are nowadays busy putting existing textbooks on CD-ROM, with hypertext links from subject to subject. In the future, such CD-ROM-based texts will also have live links to symbolic-computation software. Recently there has been a wave of speculation that instead of CD-ROM, the base for hypertext educational material should be the World-Wide Web.

Software to replace teachers and books will certainly be "multimedia", meaning it will incorporate sound and video. These features will have to be incorporated with full respect to the principle that the user is in control. Such software will not consist of videotaped 50-minute lectures. It will have live links to short video presentations, which might be either lecture-style (but limited to one short topic), or videos designed to show the applications of mathematics. Maybe you could click on the double-angle formulas for trigonometry and get a video lecture deriving them, and click again and get a video showing how they are used to design airplane wings or engines. A click in the right place might bring you a short history of those formulas, with biographical details if desired.

I have referred above to the curricular reformists who would have us emphasize concepts instead of techniques. I think this argument will fade away with the introduction of software that is capable of carrying out step-by-step solutions, thus supporting instruction in mathematical techniques. Would you like a mechanic working on your car who has been trained in the concepts of engine design? Well, of course, but she had better also know-how to service the car. Similarly, we want the engineers who design jet planes to have a thorough background in mathematical techniques, as well as concepts. I don't think there is a controversy here. I believe the development of software meeting the design criteria set forth in this paper will enable the teaching of both techniques and concepts, in an integrated curriculum that has yet to be developed. Both the design and the delivery of this curriculum will be profoundly influenced by technology.

10 Use and availability of Mathpert

Development of Mathpert began in 1985; by 1989 I was using it to teach classes at San Jose State University, but the students had no access: it was used only for classroom demonstrations. In 1990 I was awarded an ILI grant from NSF to open a student mathematics learning laboratory at San Jose State, and taught two semesters of calculus to students in that laboratory using Mathpert. Several other faculty members also used Mathpert for a few classes in that laboratory. By 1994 the calculation engine of Mathpert was nearly finished, but the interface (DOS) was five years out of date. Therefore an interface was developed using Microsoft Windows 3.1 in 1994–95. As of this writing (September 1995), Mathpert is projected to reach the market in early 1996.

References

Avitzur, R., Bachmann, O., Kajler, N. (1995): From honest to intelligent plotting. In: Levelt, A. H. M. (ed.): Proceedings of the ACM International Symposium on Symbolic and Algebraic Computation (ISSAC '95), Montreal, Canada. Association for Computing Machinery, New York, pp. 32–41.

Beeson, M. (1989): Logic and computation in Mathpert: an expert system for learning mathematics. In: Kaltofen, E., Watt, S. M. (eds.): Computers and mathematics. Springer, Berlin Heidelberg New York Tokyo, pp. 202–214.

Beeson, M. (1990): Mathpert: a computerized environment for learning algebra, trig. and calculus. J. Artif. Intell. Educ. 2: 1–11.

Beeson, M. (1992): Mathpert: computer support for learning algebra, trigonometry, and calculus. In: Voronkov, A. (ed.): Logic programming and automated reasoning. Springer, Berlin Heidelberg New York Tokyo, pp. 453–456 (Lecture notes in computer science, vol. 624).

Beeson, M. (1995): Using nonstandard analysis to verify the correctness of computations. Int. J. Found. Comput. Sci. 6: 299–338.

Brockschmidt, K. (1994): Inside OLE 2. Microsoft Press, Redmond, WA.

Buchberger, B. (1990): Should students learn integration rules? ACM SIGSAM Bull. 24: 10–17.

Bundy, A. (1983): The computer modelling of mathematical reasoning. Academic Press, London.

Burton, R. R. (1982): Diagnosing bugs in a simple procedural skill. In: Sleeman, D. H., Brown, J. S. (ed.): Intelligent tutoring systems. Academic Press, London, pp. 157–185.

Clarke, E., Zhao, X. (1992): Analytica: an experiment in combining theorem proving and symbolic manipulation. Technical Report CMU-CS-92-1 7, School of Computer Science, Carnegie Mellon University, Pittsburgh, PA.

Fateman, R. (1992): Honest plotting, global extrema, and interval arithmetic. In: Wang, P. S. (ed.): Proceedings of the ACM International Symposium on Symbolic and Algebraic Computation (ISSAC '92), Berkeley, California. Association for Computing Machinery, New York, pp. 216–223.

Nguyen-Xuan, F., Nicaud, J. F., Gelis, J. M., Joly, F. (1993): Automatic diagnosis of the student's knowledge state in the learning of algebraic problem solving. In Brna, P., Ohlsson, S., Pain, H. (eds.): Artificial intelligence in education. Association for the Advancement of Computing in Education, Charlottesville, VA, pp. 489–496.

Nicaud, J. F. (1994): Building ITSs to be used: lessons learned from the APLUSIX project. In: Lewis, R., Mendelsohn, P. (eds.): Lessons from learning. North-Holland, Amsterdam, pp. 181–198 (IFIP transactions, series A, vol. 46).

Nicaud, J. F., Gelis, J. M., Saidi, M. (1993): A framework for learning polynomial factoring with new technologies. In: International Conference on Computers in Education 93, Taiwan, pp. 288–293.

Richardson, D. (1968): Some unsolvable problems involving elementary functions of a real variable. J. Symb. Logic 33: 511–520.

Wu, W.-T. (1986): Basic principles of mechanical theorem-proving in elementary geometries. J. Automat. Reason. 2: 221–252.

Hypermedia learning environment for mathematical sciences

Seppo Pohjolainen, Jari Multisilta, and Kostadin Antchev

1 Introduction

Computers play an essential role in research and education in applied mathematics and the natural and technical sciences. Graphical interfaces have made it easier to use computers, so that nowadays many educational and research problems can be conveniently solved with existing mathematical software and hardware. Graphical object-oriented programming environments such as HyperCard (Apple Computer 1987) and ToolBook (Asymetrix 1991) have made it possible to easily integrate text, graphics, animations, mathematical programs, digitized videos and sound into hypermedia (Ambron and Hooper 1990, Jonassen and Mandl 1990, Jonassen and Grabinger 1990, Kalaja et al. 1991, Nielsen 1990). Typically, hypermedia programs contain large amounts of data. Fortunately, these can be put on CD-ROMs.

In this chapter, the design and implementation of a hypermedia-based learning-environment software, called HMLE, will be discussed. HMLE was used to implement a hypermedia-based course on matrix algebra at the Department of Mathematics at Tampere University of Technology (Multisilta and Pohjolainen 1993a, b; Pohjolainen et al. 1993a, b). Lecture notes, exercises, and examples (Pohjolainen 1993) have been written as hypertext and links to Matlab and other mathematical programs have been made in order to make possible numerical experiments and graphics. The role of videos, sound, and interactive graphics will be discussed in forthcoming papers. The main goals of the project are:

- to create a hypermedia-based learning environment for the mathematical sciences that fully integrates hypertext, computer-aided exercises, graphics, videos, and sound;
- to use this environment to develop a pilot hypermedia course in matrix algebra using existing material (lecture notes, Matlab exercises, and Mathematica notebooks);
- to support self-study by having all the material prepared for personal computers and distributed on CD-ROMs (the user may print lecture notes when needed);
- in order to minimize the cost, to make the plain hypertext readable with no need for commercial software (however, commercial software should be easily integratable into the learning environment if available);
- to make authoring, maintenance, and updating of the material as easy as possible by automatic link generation and implicit linking, for example;
- to record to some extent the actions of the learner in order to examine different study styles (this will serve as a basis for a computer-aided evaluation and advice system in the future).

The main achievement is not only a single hypermedia course in mathematics, but also a set of software tools for translating other lecture notes in the mathematical sciences into hypermedia. These tools will be used in the future to create a hypermedia-based mathematics training course for students starting their studies at university.

An introduction to the main issues to be considered when creating hypermedia books is given by Cohen and Meertens in their contribution to this volume. In this chapter we shall try to present some solutions suitable for Macintosh and PCs.

In Sect. 2 some existing hypermedia-based learning environments for mathematics are briefly reviewed. In Sect. 3 the key components of mathematical hypermedia are introduced and their role in the design of hypermedia-based learning environments is discussed. Section 4 presents the main design principles of HMLE. Section 5 describes the structure of the HMLE database: how to divide existing material into nodes and the types of nodes that can be in HMLE.

Section 6 describes the learning interface of HMLE and introduces various learning strategies that HMLE supports. The available tools include concept maps and computer-based interactive exercises.

Section 7 outlines the conclusions and the research topics that the authors have found to be important for the future versions of HMLE.

2 Related work in the field

A considerable amount of computer-based educational software on mathematics is available. The purpose of many experimental applications is to teach and exercise just one topic (for example matrix product). However, there are also complete courses implemented in computer form. For instance, "Calculus & Mathematica" is a calculus course that uses Mathematica notebooks to present theory, examples, and exercises (Davis et al. 1994). Students can fully access Mathematica by typing in formulas in Mathematica command language. Each notebook introduces the theory of the topic, which is followed by examples.

The Transitional Mathematics Project at Imperial College London has produced Mathematica notebooks to be used in calculus courses (Kent et al. 1995). At the end of each notebook there is a set of exercises to be solved with pen and paper and evaluated by the computer. Some feedback is provided for the student ("you got 2 out of 4 correct").

This kind of material certainly has its place in the classroom. It offers the capabilities of computer algebra systems (CAS) to students right from the beginning of their mathematics studies. One of the disadvantages with CAS, pointed out also by Avitzur in his contribution to this volume, is that students must learn a specific command language before they can study mathematics. Also in most cases the hypertext capabilities of CAS are limited. Apart from the possibility to open and close notebook cells, there are no hypertext features in Mathematica 2.2.

Until recently only a few courses in mathematics have been implemented as hypermedia. One example is "A Simple Introduction to Numerical Analysis" (Harding and Quinney 1992). The material is published on a CD-ROM that contains the hypertext material, animations and graphical tools. Another interesting course is a hypertext version of "Introduction to Algorithms" (Cormen et al. 1990). This CD-ROM contains hypertext, several animations, and QuickTime movies. These high-quality texts and animations add a new dimension to learning mathematics. To see a sorting algorithm in action gives the student a mental reference or orientation basis with which to relate the theory and implementation of that algorithm.

However, the authors feel that there is a problem related to the way in which the material is put together. For the reader the material seems to be a collection of "screens", "cards", or "stacks" that does not necessarily form an entity. This drawback is partly due to a limitation in HyperCard that has been used in implementing "Introduction to Algorithms": a stack typically corresponds to an application and cards in the stack are displayed only in one window. As a result, when a new card is turned on, all the previous information disappears with the old card.

3 Components of mathematical hypermedia

Hypertext in this context means a database in which information has been organised nonsequentially or nonlinearly (Conklin 1987). The database consists of nodes and links between nodes. There can be many links from a node to other nodes, and a node can be referenced by many links as well. A link is defined by source and destination nodes, and by an anchor in these nodes (Haan et al. 1992, Nielsen 1990). The destination of a link can be a file (so-called string-to-lexia link) or a string in a file (string-to-string link) (Russell and Landow 1993). With a string-to-lexia link it is not possible to reference to a certain part of a file (Fig. 1). This kind of link can make hypermedia easily navigable, especially if the destination nodes are "short" documents. String-to-string links would permit the destination to be a string in a file, but this kind of link requires more planning in the design process. *Implicit links* are generated by the hypermedia software on runtime, for example HMLE generates referential links from the mathematical concepts to their definition files. To some extent hypermedia software should generate links from the fluctuated forms of words (a link from the words "matrix" and "matrices" to the same definition file). In contrast to implicit links, *explicit links* are generated by the hypermedia author.

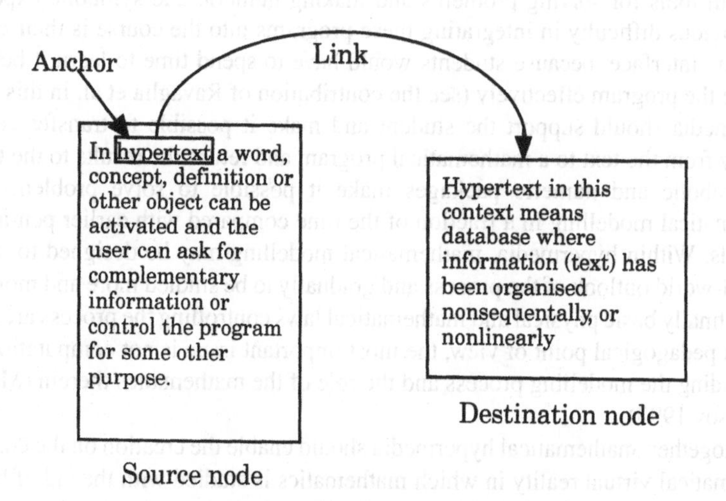

Fig. 1. Definition of string-to-lexia link

The nodes and links form a network structure in the database. Hypermedia is a database which contains pictures, digitized videos, sound, and animations in addition to text.

In hypertext, a word, concept, definition, or other object can be activated and the user can ask for complementary information or control the program for some other purpose. If the provided explanation is not sufficient, it may be completed. Mathematical text consists typically of definitions, axioms, and theorems which may be deduced from the axioms. The structure of mathematics suggests the use of string-to-lexia links, implicit linking and typed nodes. Hypertext provides a good database for modelling the structure of mathematics and mathematical courses. The basic form of mathematical hypertext is an electronic book with a dictionary of definitions, which may be accessed and studied when needed.

The main advantage of hypertext from the user's point of view is that it is able to adapt to the user's needs. Since everybody need not read all the material, hypertext supports reading and studying in different ways and on different levels, depending on the user. For the author this means that he has to write the text to support reading on different levels. This is not always so difficult, since hypertext may organise the writing process, too.

Visual information is often easy to understand. The traditional, literal way of presenting mathematics is based on formulas and text. Many essential ideas in mathematics are geometrical or visual by nature and they can be shown and studied by computer graphics (Stowasser 1984, 1991). Using a mouse, graphical elements can be made interactive so that the student may in a sense move in 3- or even n-dimensional space to get a better understanding of functions, surfaces, planes, and similar objects. Animation helps to understand time-dependent behavior, such as the solution of ordinary differential equations, two-dimensional partial differential equations, and iteration procedures in numerical analysis, to name a few examples.

Mathematical software such as Matlab (The MathWorks 1989) in matrix algebra and numerical computations, Mathematica (Wolfram 1988) and Maple (Char et al. 1991) in symbolic algebra, and general programs like MathCAD (Mathsoft 1993) provide powerful tools for solving problems and making numeric and symbolic experiments. One obvious difficulty in integrating these programs into the course is their command language interface, because students would have to spend time to learn it before they can use the program effectively (see the contribution of Ravaglia et al. in this volume). Hypermedia should support the student and make it possible to transfer commands directly from the text to a mathematical program and return the results to the text.

Symbolic and numeric packages make it possible to solve problems such as mathematical modelling in a fraction of the time compared with earlier pen-and-paper methods. Within hypermedia, mathematical modelling may be designed to start from the real-world outlook of the process and gradually to be studied more and more deeply, so that finally basic physical and mathematical laws controlling the process are revealed. From a pedagogical point of view, the most important issue is not computation but understanding the modelling process and the role of the mathematics therein (Moscardini and Ersoy 1993).

Altogether, mathematical hypermedia should enable the creation on the computer of mathematical virtual reality in which mathematics is studied with the aid of hypertext, graphics, animation, digitized videos, etc. In this context the student's role can be active;

he or she can experiment and learn by doing. As such, mathematical hypermedia supports self-study, by giving all these possibilities to the users and by instructing and evaluating them.

4 Designing mathematical hypermedia

In general, the design of a learning environment can be divided into three tasks: modelling the database (information model), modelling the user interface, and modelling the learning interface (Jonassen and Grabinger 1990). In Sects. 5 and 6 the design of the HMLE for matrix algebra is discussed from these three points of view. However, in this application the user is also the learner and hence it is more natural to present the user interface at the same time as the learning interface is discussed.

In this section conversion and presentation problems of mathematical text and formulas in simple hypermedia systems will be discussed. See the contribution of Quint et al. in this volume, for a more detailed presentation.

The implementation of HMLE was carried out in the Macintosh environment, because there was a free hypermedia application (HyperCard), the Macintosh system software supported videos (QuickTime), and existing lecture material was written with Microsoft Word on a Macintosh. More details on the architecture of HMLE can be found in the final section.

4.1 From lecture notes to hypertext

The design method and software presented here are intended to be used in designing a hypermedia course based on lecture notes available in computer form. The main goal is to minimize the work for the author by relying heavily on existing lecture notes. This approach for developing hypermedia is discussed by Rada (1992), where conversion is performed in two steps. *First-order hypertext* reflects the original markup of the document. *Second-order hypertext* adds links that are not explicit in the text. Rada (1992) marked the documents with SGML (Standard Generalised Markup Language). Our documents use a different markup, namely RTF (Rich Text Format) (Microsoft 1993), but in principle the conversion process is similar to Rada's (1992). RTF contains all the text formatting information, pictures, and formulas and it can be transferred between different word processors or between Macintosh and PCs.

The major difference between SGML and RTF is that SGML describes the structure of a document, whereas RTF describes the physical characteristics of the text (text face, size, etc.). However, RTF also includes certain tags that describe document structure. The author can define a set of styles for the document (heading 1, heading 3, abstract, etc.) that are written into the beginning of the RTF file and have a special tag in the RTF markup.

4.2 Presentation of mathematical texts

The presentation of mathematical formulas is a common problem for all educational software. For example, Mathematica 2.2 displays formulas using a monospaced font (for example Courier) so that Mathematica is able to print formulas on ASCII terminals as well as on notebooks. Calculus & Mathematica (Davis et al. 1994) contains a special font to improve the quality of formulas. Kent et al. (1995), created formulas with a formula editor and transferred them to notebooks as pictures. In the future, formulas should be active parts of documents as suggested by Quint et al. in this volume.

To make authoring easy and to improve readability, mathematical text should be written with a text editor that is able to display formulas sufficiently well. Our selection was Microsoft Word 5.x, which includes an equation editor. Other alternatives were FrameMaker (Frame Technology 1990) and TEX (Knuth 1984). The advantage of Microsoft Word for our project was that lecture notes at the Department of Mathematics at Tampere University of Technology had already been written with it. If needed, the equation editor in Word can also generate formulas in TEX format.

Microsoft Word, FrameMaker, and TEX are not hypermedia systems as such. Actually, FrameMaker 4 has basic hypertext functionality in that it allows simple linking. However, it does not fully integrate other programs, sound, and videos as a part of the text. In general, it is easier to program full-scale hypermedia using a hypermedia development environment that is easily extendible by writing new code modules.

Apple Macintosh has published HyperCard (Apple Computer 1987), an object-oriented programming environment, with buttons, fields, and links to other programs, which can be quite easily programmed (Goodman 1990). Graphics, animation, and sounds can be presented with HyperCard and QuickTime system extension. In addition, HyperCard can be extended with external commands or functions (XCMDs) that are written in C or Pascal (Apple Computer 1992). ToolBook (Asymetrix 1991) is an equivalent software for Windows that is widely available for PCs.

HyperCard can display text in so called text fields. A text field is a transparent layer in front of the corresponding card. Unfortunately, HyperCard is not able to display mathematical formulas in text fields. This is why formulas must be displayed on cards as pictures. This makes authoring difficult, because the author has to place the formulas separately on the card to correspond to the related text. Moreover, the contents of the card cannot be scrolled while text fields are scrollable.

In order to improve the handling of mathematical formulas in HyperCard we designed and implemented an XCMD, called RTF Reader that can display mathematical text with formulas in a scrollable window. RTF Reader can read and display RTF files and so called *converted files*. RTF files are prepared with Microsoft Word 5 and they are converted to the converted file format by RTF Converter. The converted files are faster to open and use less disk space. Every picture, equation, and word in RTF or converted file can be activated and linked to other nodes.

4.3 Architecture of HMLE

The HMLE developed in this work consists of several subsystems. The central element is a HyperCard stack that contains a table of contents for the hypermedia material and databases for links and nodes. Other subsystems are RTF Reader for presenting mathematical hypertext, Node Tool for maintaining nodes, Link Tool for maintaining links, Link, Palette for activating a link, and Concept Map Tool for producing concept maps (see Fig. 2). They are all implemented as external commands in HyperCard. There can also be several applications in HMLE. Exercise Maker (EM) is a stand-alone tool that can generate and check numerical exercises. EM can also present theoretical exercises and provide hints to the proofs. The role and implementation of EM will be discussed in Sect. 6. Matlab and Mathematica are used to execute numerical and symbolic examples

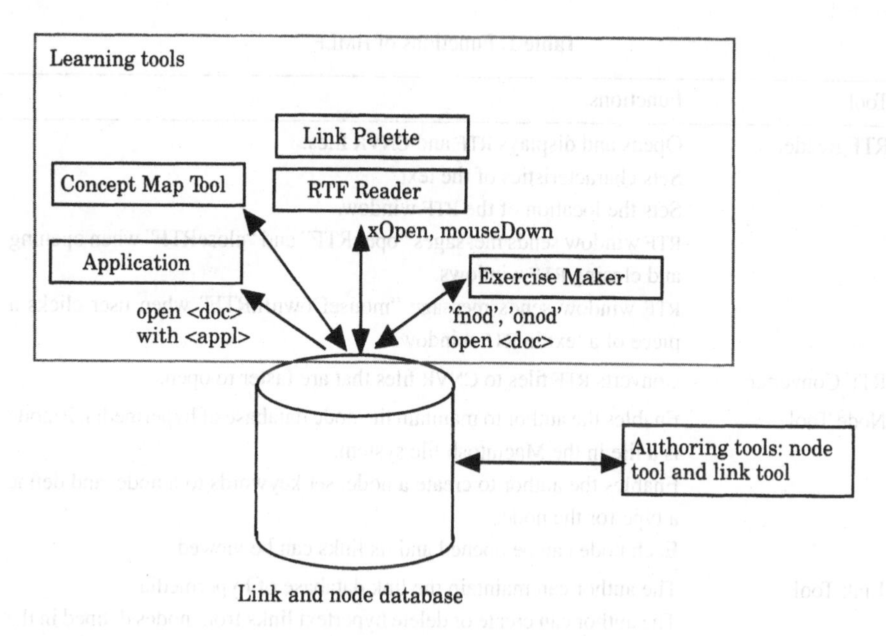

Fig. 2. Architecture of HMLE

and exercises. In addition, there can be QuickTime movies that are played by QuickTime Movie Player. The functions of the tools are presented in Table 1.

5 Structure of the hypermedia database

The HMLE database contains two types of information, namely nodes and links between nodes. The database was designed so that nodes are saved as separate files and only a reference to the actual node (file name) and some additional information are saved into the database. The other part of the database consists of the link information. It is appropriate to keep it separate from the node information and actual documents because it then becomes possible to use any file in the Macintosh as a node in HMLE.

The lecture notes can be divided into nodes in many ways. In the matrix algebra course there are 72 subchapters that contain the original lecture notes. Subchapters form a database that can be read sequentially or linearly as original lecture notes on paper.

It is important in designing the database to understand what a user is seeking when a hypertext document is opened and a word selected. For example, if users activate the word "eigenvalue", then they probably want to know the definition of the word. A straightforward way to find the definition is to open the lecture notes on the page where "eigenvalue" has been defined. In reading the definition for "eigenvalue", one may ask for explanations of other unknown concepts that have been used, or may return to the page from where he or she started. The problem is that while gathering information from different places – some people put their fingers between the pages in a book – one may have difficulties in navigating back to where one started. A better method is to write each definition of a mathematical concept in a separate file and open them in separate windows when the concept in the text has been activated. The obvious advantage is that the text

Table 1. Functions of HMLE

Tool	Functions
RTF Reader	Opens and displays RTF and CNVR files. Sets characteristics of the text. Sets the location of the RTF window. RTF window sends messages "openRTF" and "closeRTF" when opening and closing RTF windows. RTF window sends message "mouseDownInRTF" when user clicks a piece of a text in RTF window.
RTF Converter	Converts RTF files to CNVR files that are faster to open.
Node Tool	Enables the author to maintain the node database of hypermedia. A node is a file in the Macintosh file system. Enables the author to create a node, set keywords to a node, and define a type for the node. Each node can be opened and its links can be viewed.
Link Tool	The author can maintain the link database of hypermedia. The author can create or delete hypertext links from nodes defined in the Node Tool. Links from HyperCard fields to nodes and to cards of the current stack are also possible.
Link Palette	Activates its buttons when user clicks a word in RTF window.
Concept Map Tool	Draws a concept map (tree) of a specified concept. By clicking the concept in a concept map tree, it shows the definition of the concept.
Annotation Tool	Allows the user to add his or her own notes to a node.
Exercise Maker	Implements interactive sessions for HMLE. Can read exercises from RTF files divided by sections (a section contains one level of hints). Can use Mathematica through MathLink for generating exercises and checking the results.
HMLE Stack	Controls all tools by HyperTalk (see Table 2). Contains resources that other tools need (error messages, icons, node database, link database). Plays the QuickTime video. Implements the table of contents and keywords list. Controls applications needed to view some of the nodes (Matlab, Mathematica, Maple).

and all the explanations may be seen at the same time. The request for the next concept opens a new window, so that the problem of navigation is minimized. Fortunately the number of mathematical concepts in the matrix algebra course is rather small. At this moment there are about 80 separate definition files, mostly of mathematical concepts such as "singular".

Table 2. XCMD's and message handlers used by HMLE stack

XCMD or message handler	Function
GetAppName XCMD	Resolves the name of an application that can open a specified node.
GetTargetNode XCMD	Resolves the destination of a specified link.
LinkEvent message handler	Access function. Follows the link by opening the destination node with the correct application.
GetKeyWords	Updates the list of keywords by examining the node data.
FrontRTFWindow	Returns the name of the frontmost RTF window.
DoBasicForm	Converts a fluctuated form of a name of a mathematical concept (that is present in the keywords list) to its basic form.

Opening separate windows for definitions and concepts opens up an interesting freedom in writing hypertext lecture notes. The definitions given in these files may contain more information than the definitions used in the original text. It is boring to read detailed definitions if you master them well; if you do not, you will be grateful for every clue that improves your understanding. For the writer, hypertext is different from the original text. The presentation will change depending on the actions of the reader.

As the definition files are opened in separate windows, one might ask for additional information for the concepts to understand them better. That is why most concepts should be studied not only as definitions, but as examples, exercises, and in numeric and symbolic way with Matlab, Mathematica, or Maple. In most cases it is easy to give examples of selected topics simply by writing the examples in separate files that will be opened when needed. In addition to subchapters and definitions, examples and exercises are also saved as separate files and they form their own nodes. In the present form the hypertext database consists of subchapter, definition, example, exercise, tool, and stack nodes.

Definition node

The definition node contains the definition of a mathematical object. The definition can be an elementary-level definition such as a determinant, but also a higher-level concept like the LU-decomposition. Definitions are written in as complete a form as possible and can be read as hypertext so that they may be cross-referenced. In addition to the exact mathematical definitions, extensive verbal explanations are also given.

Example node

The example node contains examples of the selected topic which can be activated in hypertext. The purpose of the examples is to clarify the definitions, decompositions, and methods, e.g., in the example associated with the definition of determinant, values of determinants may be computed in different ways.

Exercise node

The exercise node provides exercises on a given topic. There are three kinds of exercises which may be studied on different levels: check questions, numerical exercises, and theoretical exercises.

The implementation of exercises in hypertext is interesting because the computer can be used both to create these exercises and to check the answers. There are several types of exercises, the easiest being simple questions which may be answered by "yes" or "no", for example, asking whether a matrix is singular or not. A second class of exercises consists of numerical exercises, where students are asked to compute something, such as eigenvalues or the singular-value decomposition of a simple matrix. After getting the answer, the computer may check the results and tell the user whether the result is correct or not, and if not, what exactly went wrong.

Numerical exercises can easily be varied since a random-number generator can be used to generate matrices. A useful ring for this purpose consists of integer matrices which can be used to clarify the basic operations (addition, multiplication) and decompositions (the LU-decomposition, the QR-decomposition, the singular-value decomposition, the Jordan canonical form), to name a few examples.

Most of the exercises are theoretical, where a student is asked to show or to prove something. Two kinds of hints are given. The first-level hint explains the question in more detail, the second-level hint reveals the structure and results needed in the proof and the student is asked to construct the whole proof step by step. Finally, complete solutions are given for some exercises. Exercise Maker (Sect. 6) supports these different types of exercises and is able to check solutions to some extent.

Tool node

The tool node contains an executable script of the specified tool program. In order to execute the script, the tool program must be installed, e.g., Mathematica or Matlab. If it is not available, then the tool node cannot be used, although other parts of hypertext are still available. There are public-domain softwares that can be used as a simple tools. MathReader can be used to view Mathematica notebooks, and QuickTime movie players can be used to execute QuickTime movies. Some of these nodes can contain examples of numerical methods or exercises which should be solved with the tool program.

An essential part of any course on applied mathematics consists of numerical and symbolic computations with existing mathematical software, such as Mathematica, Maple, MathCAD, and Matlab. An interesting question here is the integration of mathematical programs with hypertext.

Matlab. The basic data element in Matlab is a matrix that does not require dimensioning. Defining matrices is easy and elementary manipulations can be made simply by typing formulas. Matlab contains a large number of matrix algebra commands. It is also a programming language that allows the user to study numerical methods and practice mathematical programming.

In the matrix algebra course Matlab is a natural companion because it supports matrix computation at all levels and there is much software for different purposes, e.g. Hill and Zitarelli (1994), Marcus (1993), Shahian and Hassul (1993), Del Corso (1993).

Integrating a tool program with mathematical hypertext is a very important problem. From this point of view, it is useful that the latest version of Matlab, Matlab 4.2 for Macintosh, supports Apple Events that enables communication between Matlab and HMLE.

Mathematica. Mathematica provides symbolic computation facilities for HMLE. From the design point of view, symbolic-algebra packages should be integrated so that they are easy to use, and possibly run in the background, because their user interfaces are not always well suited for educational purposes. Applications can easily communicate with Mathematica through MathLink (Wolfram Research 1992). HyperMathLink is a HyperCard interface to Mathematica that uses the MathLink protocol (Rogers and Stein 1993). The user can write an expression containing variables. The value of a variable can then be changed using a slider in the HyperMathLink stack while algebraic results are computed background in real time. In other words, users receive answers continuously as they change the values of the variables. To be able to see how the result changes as variables are altered gives a visual clue to the nature of the solution. For example, by evaluating

```
N[MatrixForm[Inverse[{{1,2,3},{4,5,6},{7,8,7+a}}]]]
```

students can study how the inverse of a matrix behaves near singularity $a = 2$ as the function of a.

Alternatively, MathReader can be used to open and display Mathematica notebooks that contain text, Mathematica commands, graphics, and animations.

Video and animation presentations. Graphical elements, animations and video clips, may be shown as QuickTime movies (see Fig. 3). A movie can be paused and viewed frame by frame. Iteration procedures, solutions, and other time-varying features may be shown.

Fig. 3. An amusing example of animation: physical inversion of matrix in three dimensions

Stack node

The stack node can contain any information that can be presented in a HyperCard stack (graphics, sound, video segments, text). For example, links to existing educational HyperCard stacks can be added to HMLE.

Sound can be used during the animations to point out something important just as when a person is watching a live show on the screen. Because the speed and storage capacity of personal computers are rapidly increasing, it should be soon possible to record complete courses on CD-ROMs or on videodisks and use them to obtain immediate on-screen advice while studying the material.

6 Learning interface

In this section the learning interface and learning aspects of HMLE are discussed. Mathematical hypermedia places special requirements on the user interface. All the buttons and icons and graphical objects should be intuitive and form a natural part of the hypertext. Figure 4 shows a part of the hypertext course of matrix algebra in the RTF-window.

There are several different ways to read the hypertext; students need not study all the material in the linear order. Two ways to organize the reading process are given as flow diagrams (orientation basis) and concept maps, which are discussed in this section.

Finally, interactive exercises for the matrix algebra course are presented. Exercise Maker uses Mathematica throughout MathLink to generate exercises and to check answers.

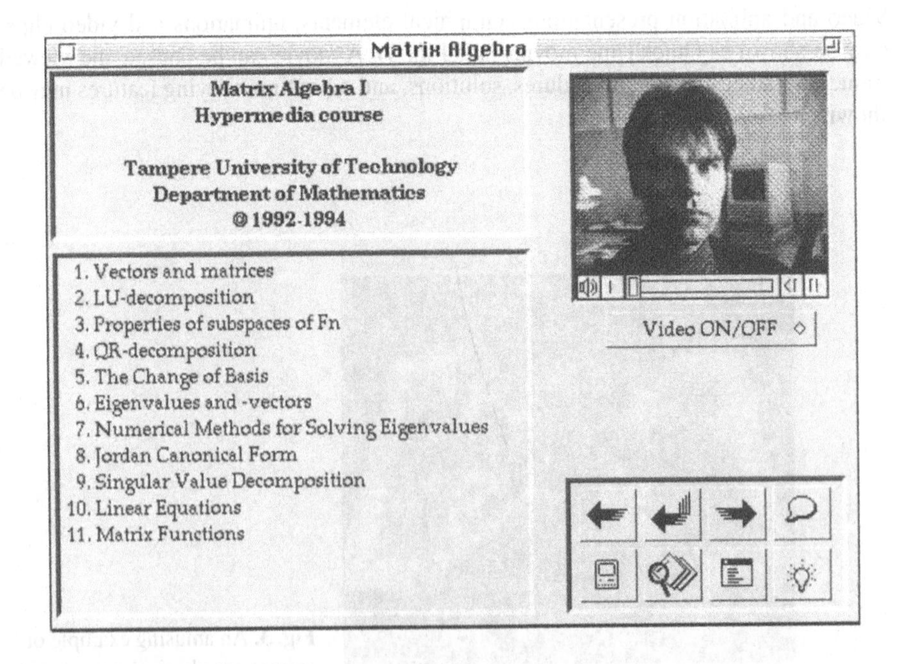

Fig. 4. Front page of the Matrix Algebra hypermedia course showing the table of contents

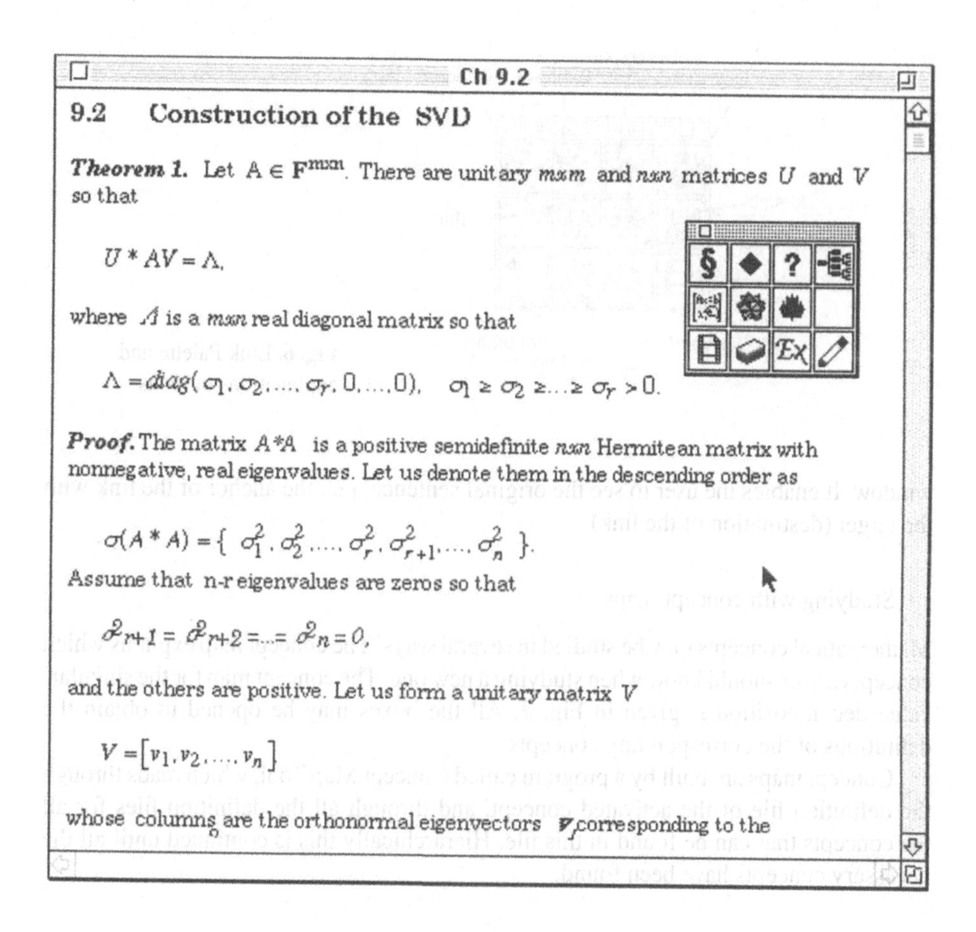

Ch 9.2

9.2 Construction of the SVD

Theorem 1. Let $A \in F^{m \times n}$. There are unitary $m \times m$ and $n \times n$ matrices U and V so that

$$U^* AV = \Lambda,$$

where Λ is a $m \times n$ real diagonal matrix so that

$$\Lambda = diag(\sigma_1, \sigma_2, \ldots, \sigma_r, 0, \ldots, 0), \quad \sigma_1 \geq \sigma_2 \geq \ldots \geq \sigma_r > 0.$$

Proof. The matrix A^*A is a positive semidefinite $n \times n$ Hermitean matrix with nonnegative, real eigenvalues. Let us denote them in the descending order as

$$\sigma(A^* A) = \{ \sigma_1^2, \sigma_2^2, \ldots, \sigma_r^2, \sigma_{r+1}^2, \ldots, \sigma_n^2 \}.$$

Assume that n-r eigenvalues are zeros so that

$$\sigma^2_{r+1} = \sigma^2_{r+2} = \ldots = \sigma^2_n = 0,$$

and the others are positive. Let us form a unitary matrix V

$$V = [v_1, v_2, \ldots, v_n]$$

whose columns are the orthonormal eigenvectors v_i corresponding to the

Fig. 5. Hypertext lecture notes on matrix algebra

Studying with hypertext

When the hypertext lecture notes are opened (Fig. 4), the student can browse through the chapters and subchapters and may try exercises and study previous examination questions. Although the hypertext can be read in any order, there is a natural order given by the table of contents (Fig. 4).

Let us select the subchapter on eigenvalues and eigenvectors. When the heading is activated, the correct subchapter will open and present the lecture notes (see Fig. 5).

Every word in the text may be selected and the user may ask for a definition, examples, exercises, and numerical work with Matlab. For example, users who are interested in eigenvalues can activate the word eigenvalues and with the Link Palette (see Fig. 6) they can get the definition, examples, numerical and theoretical exercises for the selected concept.

Although the number of open windows may become large, the authors feel that it is better to open a separate window for each node than to scroll the contents of an existing

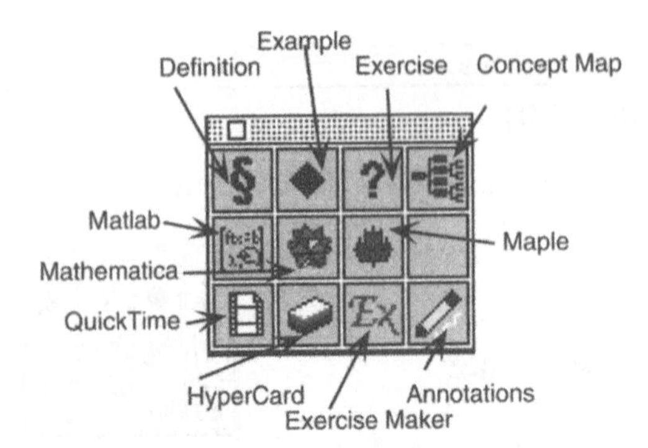

Fig. 6. Link Palette and explanations of icons

window. It enables the user to see the original sentence, i.e., the anchor of the link with the target (destination of the link).

Studying with concept maps

Mathematical concepts may be studied in several ways. The concept map explains which concepts a user should know when studying a new one. The concept map for the singular-value decomposition is given in Fig. 7. All the boxes may be opened to obtain the definitions of the corresponding concepts.

Concept maps are built by a program called Concept Map Tool, which reads through the definition file of the activated concept, and through all the definition files for all the concepts that can be found in this file. Hierarchically this is continued until all the necessary concepts have been found.

Studying with orientation basis

Some mathematical results can be given as an algorithm or computational process, and they can be presented as a flow diagram. Flow diagrams are examples of orientation

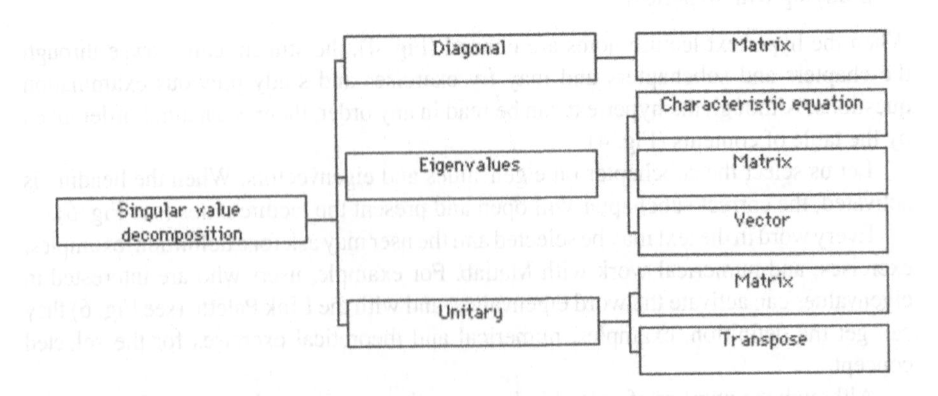

Fig. 7. Concept map for "Singular value decomposition"

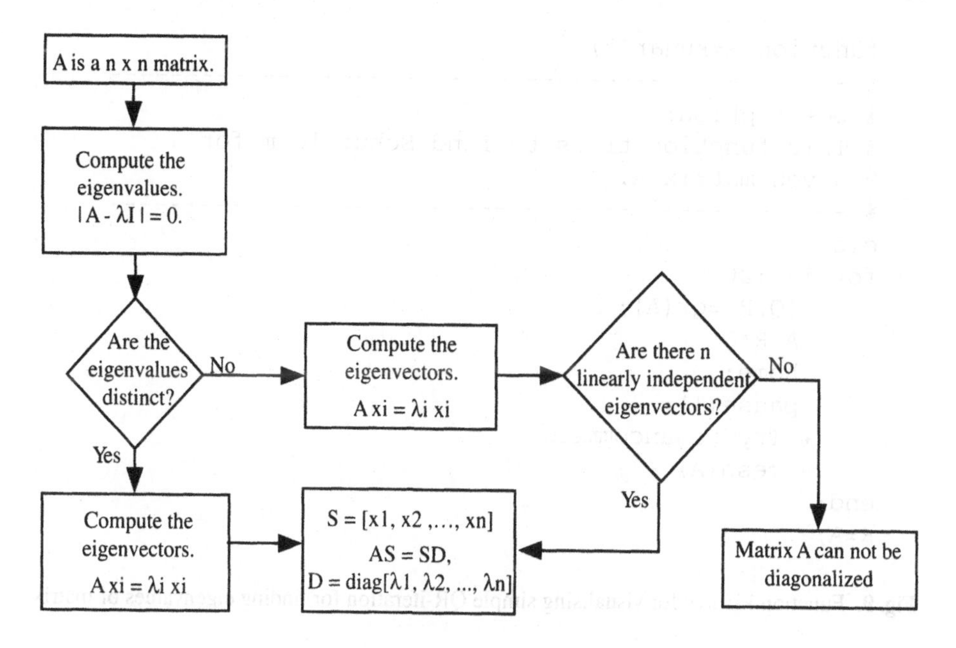

Fig. 8. Orientation basis for diagonalization of matrix

bases. Figure 8 shows an orientation basis for matrix diagonalization. A given matrix can be diagonalized with a similarity transformation along the lines given in the blocks. Every block may be opened to obtain definitions or information about what to do.

Studying with mathematical programs

Traditional pen-and-paper exercises can often be solved easily with powerful computer algebra systems and these exercises are not well suited for use in computer classes. Instead, we should endeavour to find applications and exercises where students can apply their mathematical knowledge to solve practical problems with the computer. Two exercises of this kind will be presented in this paper. In the first, a shift is introduced in the QR-algorithm to improve convergence; in the second, elementary 3-D graphics is shown (to the students) to be a straightforward application of projection operators, subspaces, and vectors. With simple Matlab programming students can understand at least two things: the power of the elementary concepts and how they help in creating 3-D graphics and animations.

QR-algorithm. The problem in this exercise is to compute the eigenvalues of a matrix by using the QR-algorithm. The student should write a short Matlab script and study the behavior of the algorithm on several random matrices. HMLE provides the student with a template for the Matlab script that contains a basic structure of the script and some comments on the implementation of the algorithm. In the final form the script `kuuar` (see Fig. 9) runs a QR-iteration which can be seen as a movie on the screen. Depending

```
function R=kuuar(A)
% ---------------------------------------------------
% Description:
% This function tries to find Schur form for a
% given matrix A.
% ---------------------------------------------------
clc
for i=1:20
     [Q,R]=qr(A);
     A=R*Q
     home;
     pause(1);
    % Try to uncomment
    % mesh(A)
end;
R=A;
```

Fig. 9. Function `kuuar` for visualising simple QR-iteration for finding eigenvalues of matrix

on the initial matrix, it is reduced to the Schur form, in which the eigenvalues of the matrix are on the diagonal, for example

```
S =
    2.3845 -0.1781  0.6435 -0.0209 -0.3121
    0.0000 -0.1114  0.8233 -0.4086  0.1951
   -0.0000  0.4359  0.0850 -0.3937 -0.3078
    0.0000  0.0000  0.0000 -0.0787 -0.4509
   -0.0000 -0.0000 -0.0000  0.0452  0.0982
```

Blocks associated with eigenvalues of almost equal modulus may arise during the iteration and deteriorate, or slow down the speed of the algorithm, so that the Schur form is not obtained (Ciarlet 1989). This is the case with the above matrix, where two blocks may be seen. At this stage the student will be asked as a second exercise to add shifts to the algorithm to overcome this drawback. As a final result, an improved QR-algorithm is obtained.

3-D graphics. Due to the fact that elementary 3-D graphics uses only simple linear algebra, it seems to be a natural field of applications for a course on matrix algebra. Professional 3-D graphics (Foley et al. 1991) uses planes in 4-dimensional spaces, where all the elementary operations can be given with multiplication matrices. From the educational, geometrical, and theoretical point of view, the following straightforward setup has some advantages (see, e.g., Fig. 10).

With Matlab it is relatively easy to define 3-dimensional objects mathematically. To show them on the screen, the user has to define a projection plane or a subspace and the

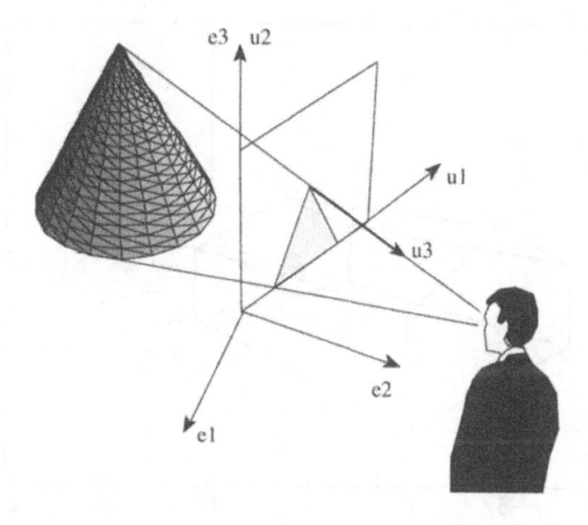

Fig. 10. Basic coordinate setting for 3-D perspective plots

direction of the projection to define parallel projections. For perspective projections the direction of the projection varies at each point.

The coordinates for a given point $z = [x_1, x_2, x_3]^T$ projected on the plane in the direction of the (column) vector u_3 are given as follows: Form a matrix

$$X = [u_1, u_2, u_3], \tag{1}$$

and let

$$Y = X^{-1} = \begin{bmatrix} y_1^T \\ y_2^T \\ y_3^T \end{bmatrix}. \tag{2}$$

Then the projection Pz of the point z in the plane spanned by the vectors u_1, u_2 is given as follows.

$$Pz = \langle y_1, z \rangle u_1 + \langle y_2, z \rangle u_2. \tag{3}$$

As a result, 3-D plots may be plotted on the subspace. An example can be seen in Fig. 11.

Studying with interactive exercises

Exercise Maker (EM) provides an interactive environment for solving exercises, checking their answers, getting advice and instructions. EM is running currently on Macintosh computers and it uses MathLink communication protocol (Wolfram Research 1992) for sending data and commands between HMLE and Mathematica kernel. It is designed in an extensible way, using an object-oriented approach. Its implementation is based on Think Class Library, an object-oriented framework for Macintosh graphic user interface (GUI) applications, part of Symantec C^{++} (Symantec Corp. 1993a, b).

Questions and exercises can be presented with the aid of EM to evaluate if the

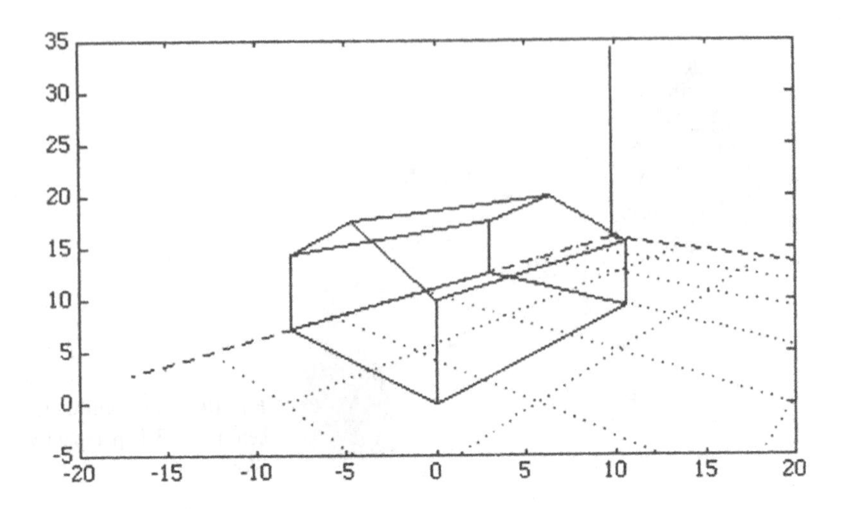

Fig. 11. Perspective projection of a house with Matlab

student has read and understood the material. Questions like "Is a non-singular matrix invertible?" may be answered simply by pushing "yes" or "no" radio controls.

Numerical exercises are useful in visualising algorithms and the structure of proofs. They will be presented with a separate exercise sheet on which the problem has been written and where students write their solution. The program informs the user as to whether the proposed solution is correct or not by comparing the answer with the correct one. The solution can easily be seen by pressing the appropriate button. Checking the results is relatively easy if the solutions are unique. Fortunately there are a lot of unique methods and decompositions in linear algebra. Another simple way to generate exercises is to start from the solution. In the case of a decomposition, for example the singular-value decomposition, the correct U, D, V matrices may be selected and the matrix A to be decomposed as the product $A = UDV^*$. To find out whether a given singular-value decomposition UDV^* is correct, it is sufficient to check that U and V are unitary matrices, that D is diagonal with nonnegative elements in the descending order, that dimensions are correct and that $UDV^* = A$.

For ease of manual computation and checking the results, calculations can be restricted to the ring of integer matrices. Luckily this ring is sufficiently rich so that most of the methods can be studied with a realistic degree of difficulty. In fact, a lot of numerical exercises with integer matrices can be generated by computer. There are results such as: an integer matrix has an integer inverse matrix if, and only if, its determinant equals plus or minus one. Computer-generated exercises are selected randomly so that the student faces different problems each time the book is opened.

In the theoretical exercises students will be asked, e.g., to show that the LU-decomposition without permutation matrix for a square nonsingular matrix is unique (Fig. 12). If the students are not able to find the solution, they may ask for hints. The first level of hints explains the question verbally in more detail and gives instructions about what to do. The second level of hints explains which results are needed to build up the

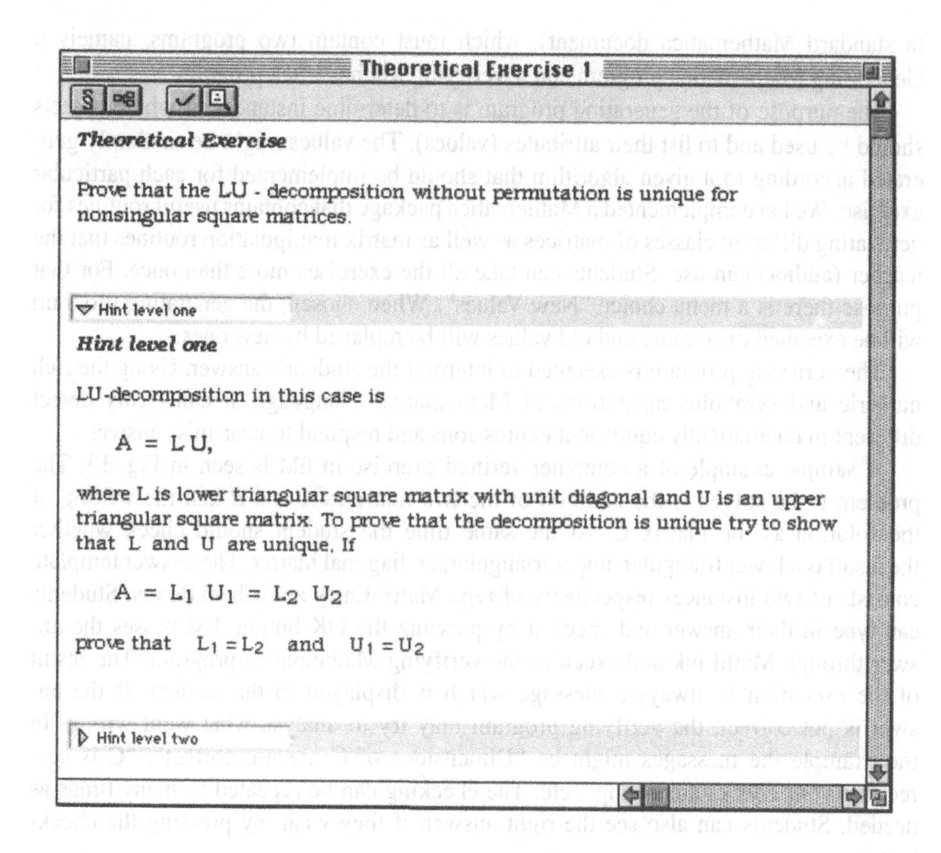

Fig. 12. Theoretical exercise

proof. The third level of hints gives the complete solution. The previous examination questions form a part of exercises and should form the most motivating class of problems for a student trying to pass an examination.

Implementation aspects of Exercise Maker. The key feature of EM is a uniform object-oriented approach to the authoring of computer-verified exercises for matrix algebra. EM defines the following set of GUI objects to be used by an author in order to construct a particular exercise: Matrix, MatrixEntry, CheckBoxes, and RadioControls.

The exercise window has two main panes: the left one is always used for stating the problem and the right one to present the answer. The answer template clearly indicates how students should enter their answer (Fig. 13). Both panes consist of instances of objects belonging to the sets mentioned above. This ensures a coherent layout for all the exercises with which students should rapidly become familiar. This also makes authoring easier.

EM uses MathLink to perform all mathematical calculations needed to generate random values for the objects in particular exercises and to interpret proposed answers. To prepare a computer verified exercise, one needs to write an ordinary notebook (a standard Mathematica document), which must contain two programs, namely a Generating Mathematica program and a Verifying Mathematica program.

The purpose of the generating program is to determine instances of which objects should be used and to list their attributes (values). The values might be randomly generated according to a given algorithm that should be implemented for each particular exercise. We have implemented a Mathematica package that contains useful routines for generating different classes of matrices as well as matrix manipulation routines that the teacher (author) can use. Students can take all the exercises more than once. For that purpose there is a menu choice "New Values". When chosen, the generating program will be executed once more and old values will be replaced by new ones.

The verifying program is executed to interpret the student's answer. Using the rich numeric and symbolic capabilities of Mathematica's language, it can easily detect different mathematically equivalent expressions and respond to near-miss answers.

A sample example of a computer-verified exercise in EM is seen in Fig. 13. The problem to be solved is the addition of the two matrices A and B and the writing of the solution as the matrix C. At the same time the student should check whether the result is a lower triangular, upper triangular, or diagonal matrix. The answer template consists of two instances respectively of type MatrixEntry and CheckBoxes. Students can type in their answer and check it by pressing the OK button. EM passes the answer through MathLink and executes the verifying Mathematica program. The result of the execution is always a message which is displayed to the student. If the answer is not correct, the verifying program may try to analyse what went wrong. In the example the messages might be "Dimensions of C are not correct", "C is correct but your checks are wrong", etc. The checking can be repeated as many times as needed. Students can also see the right answer, if they wish, by pressing the check-button.

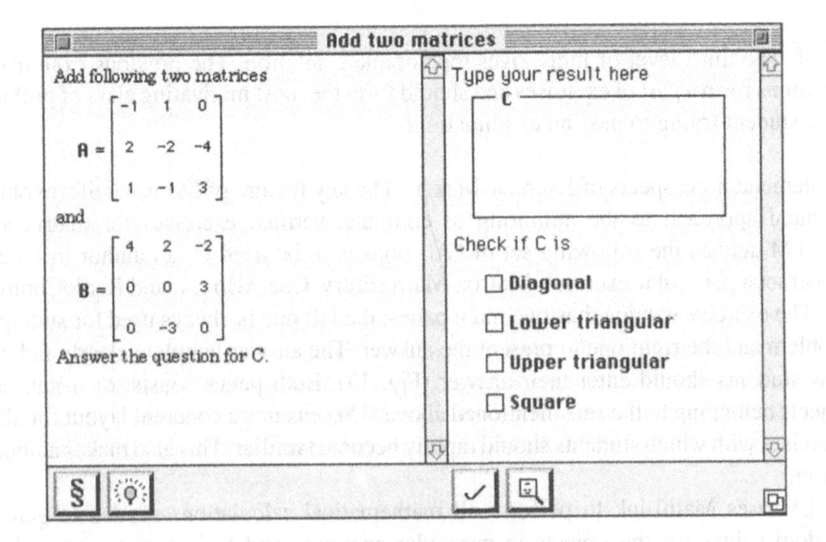

Fig. 13. Exercise opened by Exercise Maker

Exercise sheets have the hypertext property. The definition for an unknown concept may be opened by pressing the corresponding button on the toolbar of the window.

7 Conclusion and research topics

Different aspects of hypermedia that are useful in the mathematical sciences and in designing mathematical hypermedia have been discussed. A hypermedia-based learning environment for Macintosh has been designed and a hypermedia-based course on matrix algebra has been created. Further work will include testing the available pilot version in the classroom and at home. It should be thoroughly evaluated, and with the feedback from students an improved version will be designed and implemented. Also, there are many more interesting aspects to be studied. They include the following.

The integration of hypertext, exercises and examples can be improved. In the current HMLE version all these important elements are developed separately and connected in hypertext by using string-to-lexia type of links. This structure does not indicate which exercises a student should take when reading a subchapter of the text. A solution would be to create dynamically a selection of possible exercises for each subchapter or node.

Our experience on interactive exercises, where Mathematica generates exercises and checks the answers is promising. The possibilities of running mathematical programs in the background are improving all the time. Interactive symbolic and numeric exercises should be developed on all the stages of study so that students can check their skills, exercise different mathematical manipulations, and study and use simulated practical applications.

Animation, video, and sound give new possibilities for HMLE, and some works should be directed to integrate the new elements to the hypertext. Existing video clips on industrial applications, natural sciences, and other subjects can be used to demonstrate connections with real-world applications.

Some research should be directed to the nonlinear ways of learning in hypermedia environments. In HMLE a topic to be studied can be given as a high-level concept, e.g., the LU-decomposition. Students can plot the corresponding concept map and use it as a personal guide for their studies. An interesting question will arise; how much information and different learning activities should the HMLE concept boxes (Fig. 7) contain so that the original lecture notes could be replaced by the nonlinear way of learning?

The graphical look of the HMLE is still developing. It would be advantageous to have professional designers take part in the user interface design in order to improve readability, usefulness and satisfy aesthetic aspects.

Currently HMLE lacks a student model. However, it should be personalized so that it could collect and save data from the user's actions. The collected information should be used to help and guide the student and it may serve as a basis for creating a student model in the future in HMLE.

Acknowledgements

The work has been supported by Finnish Academy (Suomen Akatemia), the Finnish Ministry of Education, and Tampere University of Technology. Their support is gratefully acknowledged. The comments of an anonymous reviewer and Norbert Kajler have greatly improved this paper.

For the revision of the English manuscript we are grateful to Mr. Danny Donoghue and Mr. James Rowland.

References

Ambron, S., Hooper, K. (1990): Learning with interactive multimedia. Microsoft Press, Redmond, WA.

Apple Computer (1987): HyperCard user's guide. Apple Computer, Inc., Cupertino, CA.

Apple Computer (1992): HyperCard 2.1 release notes. Apple Computer, Inc., Cupertino, CA.

Asymetrix Corp. (1989–1991): Using ToolBook. Asymetrix Corp., Bellevue, WA.

Char, B. W., Geddes, K. O., Gonnet, G. H., Leong, B. L., Monagan, M. B., Watt, S. M. (1991): Maple V library reference manual. Springer, Berlin Heidelberg New York Tokyo.

Ciarlet, P. G. (1989): Introduction to numerical linear algebra and optimisation. Cambridge University Press, Cambridge (Cambridge texts in applied mathematics).

Conklin, J. (1987): Hypertext: an introduction and survey. IEEE Computer 20/9: 17–41.

Cormen, T., Leiserson, C., Rivest, R. (1990): Introduction to algorithms. MIT Press, Cambridge, MA.

Davis, B., Porta, H., Uhl, J. (1994): Calculus & Mathematica. Addison-Wesley, Reading, MA.

Del Corso, C. (1993): Using MATLAB in the classroom. Prentice-Hall, Englewood Cliffs (The MATLAB Curriculum series).

Foley, J. D., van Dam, A., Feiner, S. K., Huges, J. F. (1991): Computer graphics: principles and practice. Addison-Wesley, Reading, MA.

Frame Technology Company (1990): Using FrameMaker. Frame Technology Company, San Jose, CA.

Goodman, D. (1990): The complete HyperCard 2.0 handbook. Bantam Books, New York.

Haan, B. J., Kahn, P., Riley, V. A., Coombs, J. H., Meyrowitz, N. K. (1992): Iris hypermedia services. Commun. ACM 35/1: 36–51.

Harding, R., Quinney, D. (1992): MATHEMATICS, vol. 2 (CD-ROM). Anglia Polytechnic University, Renaissance Initiative.

Hill, D., Zitarelli, D. (1994): Linear algebra labs with MATLAB. Macmillan, New York.

Jonassen, D. H., Grabinger, R. S. (1990): Problems and issues in designing hypertext/hypermedia for learning. In: Jonassen, D. H., Mandl, H. (eds.): Designing hypermedia for learning. Springer, Berlin Heidelberg New York Tokyo, pp. 3–25.

Jonassen, D. H., Mandl, H. (eds.) (1990): Designing hypermedia for learning. Springer, Berlin Heidelberg New York Tokyo (NATO ASI series, series F, vol. 67)

Kalaja, M., Lehtisalo, T., Hult, S., Lassila, O. (1991): Implementing an authoring tool for educational software: the hyperreader experience. In: Uronen, P. (ed.): Proceedings of the Nordic Conference on Computer Aided Higher Education, Otaniemi, Finland. Centre for Continuing Education, Helsinki University of Technology, Espoo, pp. 172–178.

Kent, P., Ramsden, P., Wood, J. (1995): Mathematica for valuable and viable computer-based learning. In: Keränen, V., Mitic., P. (eds.): Mathematics with vision, Proceedings of the First International Mathematica Symposium. Computational Mechanics Publications, Southampton, pp. 251–258.

Knuth, D. E. (1984): The TeXbook. Addison-Wesley, Reading, MA.

Marcus, M. (1993): Matrices and Matlab: a tutorial. Prentice-Hall, Englewood Cliffs.

Mathsoft (1993): MathCAD 4.0. MathSoft Inc., Cambridge, MA.

Microsoft (1993): Rich Text Format specification. Microsoft Corp., Redmond, WA.

Moscardini, A., Ersoy, Y. (1993): Teaching mathematical modelling. In: Proceedings of the NATO Advanced Research Workshop on Mathematical Modelling in Engineering Education, Izmir, Turkey.

Multisilta, J., Pohjolainen, S. (1993a): Experiencies of hypermedia in engineering mathematics. In: Proceedings of 4th Nordic Conference under Nordic Forum for Computer Aided Education, Aalborg, Denmark, pp. 178–182.

Multisilta, J., Pohjolainen, S. (1993b): Teaching engineering mathematics with hypermedia. In: Proceedings of Technology in Mathematics Teaching, University of Birmingham, Birmingham, UK, pp. 547–00.

Nielsen, J. (1990): Hypertext and hypermedia. Academic Press, San Diego, CA.

Pohjolainen, S. (1993): Matriisilaskenta I (matrix algebra). Lecture, Tampere University of Technology, Tampere, Finland.

Pohjolainen, S., Multisilta, J., Antchev, K. (1993a): Examples of Matlab in the engineering education. In: Proceedings of NATO Advanced Research Workshop on Mathematical Modelling in Engineering Education, Izmir, Turkey.

Pohjolainen, S., Multisilta, J., Hämäläinen, T. (1993b): Tietokoneavusteisen opetuksen mahdollisuuksista. In: Neittaanmäki, P. (ed.): Matematiikan opetus. University of Jyväskylä, Department of Mathematics, Jyväskylä, Finland, Lecture Notes 24: 33–50.

Rada, R. (1992): Converting a textbook to hypertext. ACM Trans. Inf. Syst. 10: 294–315.

Rogers, C., Stein, D. (1993): HyperMathLink: a HyperCard XFCN. Wolfram Research Inc., Champaign, IL.

Russell, D. M., Landow, G. P. (1993): Educational uses of hypermedia: from design to the classroom. In: Proceedings of ACM Hypertext '93, unpublished program (courses series), Edinburgh.

Shahian, B., Hassul, M. (1993): Control system design using Matlab. Prentice-Hall, Englewood Cliffs.

Stowasser, R. J. K. (1984): Organizing ideas in the focus of mathematics for all. Science and Technology Education, UNESCO Document Series no. 20.

Stowasser, R. J. K. (1991): Beispiele zum computerorientierten Mathematikunterricht mit Logo. Technische Universität Berlin, Berlin.

Symantec Corp. (1993a): THINK Class Library guide. Symantec Corp., Cupertino, CA.

Symantec Corp. (1993b): THINK C/Symantec C^{++} user's guide. Symantec Corp., Cupertino, CA.

The MathWorks Inc. (1989): MATLAB for Macintosh computers. The MathWorks Inc.

Wolfram, S. (1988): Mathematica: a system for doing mathematics by computer. Addison-Wesley, Redwood City, CA.

Wolfram Research Inc. (1992): MathLink reference guide. Wolfram Research Inc., Champaign, IL, Technical Report.

Chains of recurrences for functions of two variables and their application to surface plotting

Olaf Bachmann

1 Introduction

When generating curves or surfaces of closed-form mathematical functions, usually the most time-consuming task is function evaluation at discrete points. Most programs (among them most of the existing computer algebra systems) achieve this by straightforward evaluations of linearly sampled points through whatever numerical evaluation routines the particular system provides. More specifically, most programs use evaluations of the following form:

$$G(x_0 + nh_x,\ y_0 + mh_y) \quad \text{for all } n = 0, \ldots, N,\ m = 0, \ldots, M$$

for some given two-dimensional function $G(x, y)$, starting points x_0, y_0 and increments h_x, h_y. For example, the following loop is used inside Maple's `plot3d` function (Char et al. 1988):

```
xinc := (xmax-xmin)/m; yinc := (ymax-ymin)/n; x := xmin;
for i from 0 to m do
     y := ymin;
     for j from 0 to n do  z[i,j] := f(x,y); y := y+yinc od;
     x := x+xinc
od;
```

Unfortunately, such general and unoptimized evaluations usually result in inefficient and long computations. Most of the common optimizations, like adaptive plotting techniques and precompilations of the given functions, seem to be of only limited help (efficient adaptive techniques for 3-D parametric surface plots are difficult to implement; compilation still usually takes quite a lot of time). A detailed discussion of the reasons for this is beyond the scope of this paper.

The results and timings presented in this paper indicate that chains of recurrences (CRs) are an effective technique to optimize the computation of *explicit* curves and surfaces. By explicit we mean curves defined by $x = f_x(u)$, $y = f_y(u)$, $z = f_z(u)$ and surfaces defined by $x = f_x(u, v)$, $y = f_y(u, v)$, $z = f_z(u, v)$, where f_x, f_y, f_z are closed-form mathematical expressions composed of analytic functions like the rational operations and the exponential, logarithmic, and trigonometric functions. The speedups gained by this method can be substantial. For example, our Maxima implementation of the CR technique needs only 0.1 s to compute the 7500 points defining an Enneper surface (see appendix). On the same machine, Maple's `plot3d` function needs 3.4 s, and a similar, straightforward Maxima evaluation even needs 60 s to compute the same Enneper surface.

The main idea behind CRs is the following: say we are given a closed-form function which we wish to evaluate at a predetermined number of equally spaced points. Instead of starting a new function evaluation from scratch at each given point, we attempt to reuse some results obtained from evaluations at previous points. That is, using the fact that the points of the evaluation range are linearly dependent, we algebraically try to deduce recurrence relations of the corresponding function values and use CRs to formalize and represent those recurrence properties.

A CR can be considered as a linearized representation of a mathematical expression which can be evaluated in a simple, vector-like fashion. The application of CRs to the computation of explicit curves and surfaces can be outlined as follows: given $f_x(u, v)$ (resp. f_y, f_z), u_0, v_0, h_u, and h_v, we first construct an equivalent CR-expression $\Phi_x(i, j)$ for which $f_x(u_0 + i h_u, v_0 + j h_v) = \Phi_x(i, j)$ for $i, j \in \mathbf{N}$. We then use Φ_x to compute the values of f_x over the required domain. In the worst case, an evaluation of Φ_x is just as expensive as an evaluation of f_x. However, for many common cases, CR evaluations are considerably more efficient, especially for expressions which have polynomial or trigonometric subexpressions.

Specific recurrence relations for many interesting functions have been investigated since the early days of mathematics. A principle known by numerical analysists as *forward differencing* – used by Charles Babbage in his famous Difference Engine No. 1 – is probably the earliest predecessor of the CR-method. Similar to CRs, forward differencing reduces the process of evaluating an n-th degree polynomial to just n additions after the first few steps. Knuth (1981, p. 469) also described this phenomenon as "tabulating polynomial values". The techniques considered here originated directly from Zima's (1984, 1992) work on systems of recurrent relations (SRRs). In Bachmann et al. (1994), Zima's original concepts of SRRs are reformulated in clearer terms and renamed to chain of recurrences. The CR paper gives detailed descriptions and properties of one-dimensional CRs and explicitly describes algorithms for constructing and evaluating them.

For practical applications such as surface computations, the CR technique needs to be extended to handle functions of more than one variable. That is the main topic of this paper. The theoretical concepts of CRs for two-dimensional functions are developed and it is shown in detail how the CR technique is applied to plotting of surfaces with Maxima and the graphing tool IZIC (Fournier et al. 1993). Section 2 summarizes the theoretical concepts of CRs for functions of one variable. Section 3 extends CRs to two-dimensional functions and gives algorithms for constructing and evaluating two-dimensional CR-expressions. Section 4 describes a Maxima implementation of the CR technique, discusses its interface to the IZIC graphing tool, and reports on the timings obtained. Some concluding remarks in Sect. 5 are finally followed by the IZIC drawings of the surfaces considered in this paper.

2 Chains of recurrences for functions of one variable

In this section we give a brief summary of the main concepts of CRs for functions of one variable. For more details, see Bachmann et al. (1994).

2.1 Chains of recurrences

Given $k \in \mathbf{N},\ k > 0$, a function $f_k(n)$ defined over $\mathbf{N}$, constants (or constant expressions not depending on n) $\varphi_0, \varphi_1, \ldots, \varphi_{k-1}$, and operators $\odot_1, \ldots, \odot_k \in \{+, *\}$, a CR Φ, represented by the tuple $\Phi = \{\varphi_0, \odot_1, \varphi_1, \odot_2, \varphi_2, \ldots, \odot_k, f_k\}$, is defined recursively as a function over $\mathbf{N}$ by

$$\Phi(n) = \begin{cases} \varphi_0 + \sum_{j=0}^{n-1} f_1(j) & \text{if } k = 1 \text{ and } \odot_1 = +, \\ \varphi_0 * \prod_{j=0}^{n-1} f_1(j) & \text{if } k = 1 \text{ and } \odot_1 = *, \\ \{\varphi_0, \odot_1, \{\varphi_1, \odot_2, \varphi_2, \ldots, \odot_k, f_k\}\}(n) & \text{if } k > 1. \end{cases} \tag{1}$$

We call $\varphi_0, \ldots, \varphi_{k-1}, f_k$ the components of the CR Φ, $k = L(\Phi)$ the *length* of Φ, Φ a *simple* CR if f_k is a constant, Φ a *polynomial* CR if Φ is simple and $\odot_1 = \odot_2 = \ldots = \odot_k = +$, Φ an *exponential* CR if Φ is simple and $\odot_1 = \odot_2 = \ldots = \odot_k = *$.

For increased readability we will consistently denote CRs by uppercase Greek letters (like Φ and Ψ) and constant CR-components by indexed lowercase Greek letters (like φ_0 and ψ_1). The shorthand notation $\Phi = \Phi(n)$ will also be used wherever unambiguous and convenient.

CRs enjoy the important property that their value for consecutive n can be computed in a very simple and efficient way based on the following observation: Given a CR $\Phi = \{\varphi_0, \odot_1, \varphi_1, \ldots, \odot_k, f_k\}$ there exist functions $f_0, f_1, \ldots, f_{k-1}$ such that for $0 \le j < k$

$$f_j(n) = \begin{cases} \varphi_j & \text{if } n = 0, \\ f_j(n-1) \odot_{j+1} f_{j+1}(n-1) & \text{if } n > 0, \end{cases} \tag{2}$$

and $f_0(n) = \Phi(n)$.

In other words, only $L(\Phi) = k$ arithmetic operations are needed in order to compute the value of a simple CR Φ at one point (assuming that the intermediate values of the functions f_j are stored and updated appropriately).

2.2 Some questions and some answers

In the light of CRs being functions defined in a peculiar way, some questions might immediately arise: What kind of "interesting", closed-form functions G can be represented by a CR? Given a suitable function G, how can its respective CR representation be constructed? Is the CR evaluation of G more efficient than the evaluation of the original function?

First, we can easily observe that a polynomial CR of length k is (as the name suggests) a polynomial in n of degree k and that an exponential CR is e raised to a polynomial exponent in n. For example,

$$\begin{aligned} \{0, +, 1\}(n) &= n, \\ \{1, *, e\}(n) &= e^n, \\ \{1, +, 2, +, 2\}(n) &= n^2 + n + 1, \\ \{1, *, 2, *, 2\}(n) &= \exp\left(\frac{\log 2}{2} n^2 + \frac{\log 2}{2} n\right), \end{aligned}$$

and more general, we can show that

$$\{\varphi_0, +, \varphi_1, +, \varphi_2, +, \dots, +, \varphi_k\}(n) = \varphi_0 + \varphi_1 \binom{n}{1} + \varphi_2 \binom{n}{2} + \dots + \varphi_k \binom{n}{k},$$

$$\{\varphi_0, *, \varphi_1, *, \varphi_2, *, \dots, *, \varphi_k\}(n) = \exp\left(\log(\varphi_0) + \log(\varphi_1)\binom{n}{1} + \log(\varphi_2)\binom{n}{2} + \dots + \log(\varphi_k)\binom{n}{k}\right).$$

Conversely, if $G(x)$ is a k-th degree polynomial in x (resp. a constant raised to a k-th degree polynomial in x) then it will be shown below that we can construct a polynomial (resp. exponential) CR Φ of length k, such that $\Phi(n) = G(x_0 + n\,h)$ for all $n \geq 0$. For example,

$$\begin{aligned}
x &= \{x_0, +, h\}, \\
2^x &= \{2^{x_0}, *, 2^h\}, \\
x^2 &= \{x_0^2, +, 2hx_0 + h^2, +, 2h^2\}, \\
e^{-x^2} &= \{e^{-x_0^2}, *, e^{-2hx_0 - h^2}, *, e^{-2h^2}\}, \\
x^2 + x + 1 &= \{x_0^2 + x_0 + 1, +, h(1 + 2x_0 + h), +, 2h^2\}, \\
x^3 &= \{x_0^3 + 3hx_0^2 + 3h^2x_0 + h^3, +, 6h^2x_0 + 6h^3, +, 6h^3\}, \\
2^{x^3}/e^{x^2} &= \{(2/e)^{x_0^3 - x_0^2}, *, (2/e)^{h^3 + h^2(3x_0 - 1) + h(3x_0^2 - 2x_0)}, \\
&\qquad *, (2/e)^{6h^3 + h^2(6x_0 - 2)}, *, 2^{6h^3}\}.
\end{aligned}$$

It is obvious that a polynomial (resp. exponential) CR of length k can be more efficiently evaluated for consecutive points than a straightforward or Horner evaluation of a polynomial of degree k, because by using (2), the CR evaluation requires only k additions (resp. multiplications) per point.

CRs can also be used to represent and efficiently evaluate sums and products of certain functions (like factorials, see Bachmann et al. 1994). It might also be interesting to observe that a CR Φ can be considered as an encoding of a straight-line program which computes the value of Φ for successive integers (Kaltofen 1988).

In order to be able to use CRs to expedite the evaluation of more complicated functions (like functions with polynomial subexpressions or trigonometric functions) it is necessary to generalize the definition of CRs to expressions whose arguments are CRs, viz., to CR-expressions.

2.3 CR-expressions

An expression Φ is called a *CR-expression*[1] if it represents one of the following functions over $\mathbf{N}$:

---

1 In later sections of this paper we will also refer to the here defined CRs and CR-expressions as one-dimensional CRs and one-dimensional CR-expressions.

i. a constant,
ii. a CR $\{\varphi_0, \odot_1, \varphi_1, \odot_2, \ldots, \odot_k, f_k\}$ where f_k is a CR-expression,
iii. $F(\Phi_1, \ldots, \Phi_m)$, where F is a function of m arguments and $\Phi_1, \ldots, \Phi_m$ are CR-expressions.

In other words, a CR-expression is basically just like a "normal" expression, except that the leaves of a CR-expression are either a constant or a CR instead of a constant or a variable. Hence, by simply replacing occurrences of the variable x with the CR $\{x_0, +, h\}$ we can transform *any* closed-form expression $G(x)$ into an equivalent CR-expression Φ (by equivalent we mean here $\Phi(n) = G(x_0 + n\,h)$ for $n \geq 0$). For example,

$$x^2 = \{x_0, +, h\} * \{x_0, +, h\} \qquad = \{x_0^2, +, 2hx_0 + h^2 + 2h^2\},$$

$$\frac{2x}{x^2+1} = \frac{2 * \{x_0, +, h\}}{\{x_0, +, h\} * \{x_0, +, h\} + 1} = \frac{\{2\,x_0, +, 2\,h\}}{\{x_0^2 + 1, +, 2hx_0 + h^2, +, 2h^2\}},$$

$$x + e^x = \{x_0, +, h\} + e^{\{x_0, +, h\}} \qquad = \{x_0, +, h\} + \{e^{x_0}, *, e^h\},$$

$$\sin(x) = \sin(\{x_0, +, h\}) \qquad = \Im(\{e^{i\,x_0}, *, e^{i\,h}\}),$$

$$\cos(x) * e^x = \cos(\{x_0, +, h\}) * e^{\{x_0, +, h\}} \qquad = \Re(\{e^{x_0 + ix_0}, *, e^{h+ih}\})$$

where the symbols $\Im$ and $\Re$ denote the imaginary and real part of complex numbers. Notice that the CR-expressions in the third column were obtained by simplifying the respective CR-expressions of the second column through the simplification rules described in the next section.

Generalizing from the length of a CR, we furthermore define the *cost index* (CI) of a CR-expression Φ to be

$$\mathrm{CI}(\Phi) = \begin{cases} 0 & \text{if } \Phi \text{ is a constant,} \\ k + \mathrm{CI}\,(f_k) & \text{if } \Phi \equiv \{\varphi_0, \odot_1, \varphi_1, \ldots, \odot_k, f_k\}, \\ 1 + \sum_{j=1}^{m} \mathrm{CI}(\Phi_j) & \text{if } \Phi \equiv F(\Phi_1, \Phi_2, \ldots, \Phi_m).\} \end{cases}$$

Similar to the length of a CR, the cost index of a CR-expression gives an indication of its evaluation cost (it counts the number of operations needed to evaluate a CR-expression at one point; for simplicity, let us count F as one operation and assume that if $F = +$ or $F = *$ then $m = 2$) and will serve us as a "simplicity-measure" of CR-expressions.

2.4 Simplification of CR-expressions

In this section we describe simplification rules which generally make evaluations of CR-expressions more efficient. This is mainly achieved by replacing (sub-)expressions of CR-expressions by CRs, viz., by reducing the cost index of the considered CR-expressions. All simplification rules of this section are proved by using the definition of CRs and some basic arithmetic identities. See Bachmann et al. (1994), for some more details.

Lemma 1. Let $\{\varphi_0, \odot_1, \varphi_1, \odot_2, \ldots, f_k\}$, $\{\psi_0, \otimes_1, \psi_1, \otimes_2, \ldots, g_l\}$ be CRs and c be a constant. Then,

$$c + \{\varphi_0, +, \varphi_1, \odot_2, \ldots, f_k\} = \{c + \varphi_0, +, \varphi_1, \odot_2, \ldots, f_k\}, \tag{3}$$

$$\begin{aligned}\{\varphi_0, +, \varphi_1, \odot_2, \ldots, f_k\} &+ \{\psi_0, +, \psi_1, \otimes_2, \ldots, g_l\} \\ &= \{\varphi_0 + \psi_0, +, \{\varphi_1, \odot_2, \ldots, f_k\} + \{\psi_1, \otimes_2, \ldots, g_l\}\},\end{aligned} \tag{4}$$

$$c * \{\varphi_0, *, \varphi_1, \odot_2, \ldots, f_k\} = \{c * \varphi_0, *, \varphi_1, \odot_2, \ldots, f_k\}, \tag{5}$$

$$\begin{aligned}\{\varphi_0, *, \varphi_1, \odot_2, \ldots, f_k\} &* \{\psi_0, *, \psi_1, \otimes_2, \ldots, g_l\} \\ &= \{\varphi_0 * \psi_0, *, \{\varphi_1, \odot_2, \ldots, f_k\} * \{\psi_1, \otimes_2, \ldots, g_l\}\},\end{aligned} \tag{6}$$

$$c * \{\varphi_0, +, \varphi_1, \odot_2, \ldots, f_k\} = \{c * \varphi_0, +, c * \{\varphi_1, \odot_2, \ldots f_k\}\}, \tag{7}$$

$$c^{\{\varphi_0, +, \varphi_1, \odot_2, \ldots, f_k\}} = \{c^{\varphi_0}, *, c^{\{\varphi_1, \odot_2, \ldots, f_k\}}\}, \tag{8}$$

$$\{\varphi_0, *, \varphi_1, \odot_2, \ldots, f_k\}^c = \{\varphi_0^c, *, \{\varphi_1, \odot_2, \ldots, f_k\}^c\}, \tag{9}$$

$$\log_c(\{\varphi_0, *, \varphi_1, \odot_2, \ldots, f_k\}) = \{\log_c(\varphi_0), +, \log_c(\{\varphi_1, \odot_2, \ldots, f_k\})\}. \tag{10}$$

Notice that the left-hand side CR-expressions are replaced by respective CRs on the right-hand side and that $CI(LHS) \geq CI(RHS)$ for (7)–(10) and $CI(LHS) > CI(RHS)$ for (7)–(10) and $CI(LHS) > CI(RHS)$ for (3)–(6). Hence, the CRs on the right-hand side of Lemma 1 are (in the spirit of the discussion above) simpler and evaluate more efficiently than the respective left-hand side CR-expressions.

Next, we consider two properties of CR-expressions for which this is generally not true.

Propositon 1. Let $\Phi = \{\varphi_0, \odot_1, f_1\}$, $\Psi = \{\psi_0, \otimes_1, g_1\}$ be CRs. Then,

$$\{\varphi_0, +, f_1\} * \{\psi_0, +, g_1\} = \{\varphi_0\, \psi_0, +, \Phi\, g_1 + f_1\, \Psi + f_1\, g_1\},$$

$$\{\varphi_0, *, f_1\}^{\{\psi_0, +, g_1\}} = \left\{\varphi_0^{\psi_0}, *, \Phi^{g_1} * f_1^{\Psi} * f_1^{g_1}\right\}.$$

However, using this proposition we can show:

Lemma 2. Let $\{\varphi_0, +, \varphi_1, +, \ldots, +, \varphi_k\}$ and $\{\psi_0, +, \psi_1\}$ be polynomial CRs. Then,

$$\begin{aligned}\{\varphi_0, +, \varphi_1, +, \ldots, +, \varphi_k\} &* \{\psi_0, +, \psi_1\} \\ &= \{\varphi_0\psi_0, +, \widetilde{\varphi}_1, +, \ldots, \widetilde{\varphi}_k, +, (k+1)\varphi_k\psi_1\}\end{aligned} \tag{11}$$

with $\widetilde{\varphi}_j = j\,\varphi_{j-1}\,\psi_1 + (\psi_0 + j\,\psi_1)\,\varphi_j$ for $1 \leq j \leq k$. Let furthermore $m \in N$. Then,

$$\{\psi_0, +, \psi_1\}^m = \{\psi_{0m}, +, \psi_{1m}, +, \ldots, \psi_{m-1,m}, +, \psi_{mm}\} \tag{12}$$

with

$$\psi_{pq} = \begin{cases} \psi_0^q, & \text{if } p = 0, \\ q!\,\psi_1^q, & \text{if } p = q, \\ p\,\psi_1\,\psi_{p-1,q-1} + (\psi_0 + p\,\psi_1)\psi_{p,q-1}, & \text{if } 0 < p < q. \end{cases}$$

And more general, we can show by induction:

Lemma 3. Let $\{\varphi_0, \odot, \varphi_1, \odot, \ldots, \odot, \varphi_k\}$ and $\{\psi_0, \otimes, \psi_1, \otimes, \ldots, \otimes, \psi_l\}$ be polynomial (resp. exponential) CRs. Then,

$$\{\varphi_0, +, \varphi_1, +, \ldots, \varphi_k\} * \{\psi_0, +, \psi_1, + \ldots, \psi_l\} = \{\varphi_0\psi_0, +, \xi_1, +, \xi_2, +, \ldots, +, \xi_{k+l}\} \tag{13}$$

$$\{\varphi_0, *, \varphi_1, *, \ldots, \varphi_k\}^{\{\psi_0,+,\psi_1,+,\ldots,\psi_l\}} = \{\varphi_0^{\psi_0}, *, \zeta_1, *, \zeta_2, *, \ldots, *, \zeta_{l+k}\} \tag{14}$$

where the constants ξ_j, ζ_j are obtained by appropriate recursive applications of (3)–(11).

Finally, we consider some simplification properties of CR-expressions involving trigonometric functions, hereby extending the results about algebraic properties of CRs given in Bachmann et al. (1994).

Proposition 2. Let $\{\varphi_0, +, \varphi_1, \odot_2, \ldots, \odot_k, f_k\}$ be a CR with $\varphi_0, \ldots, \varphi_{k-1} \in \boldsymbol{R}$, and $f_k\colon \boldsymbol{N} \Rightarrow \boldsymbol{P}$. Then,

$$\sin(\{\varphi_0, +, \varphi_1, \odot_2, \ldots, \odot_k, f_k\}) = \Im(\{e^{i\varphi_0}, *, e^{i\{\varphi_1\odot_2,\ldots,\odot_k,f_k\}}\}),$$
$$\cos(\{\varphi_0 + \varphi_1, \odot_2, \ldots, \odot_k, f_k\}) = \Re(\{e^{i\varphi_0}, *, e^{i\{\varphi_1\odot_2,\ldots,\odot_k,f_k\}}\}).$$

On a first sight, these might not look like simplifying properties and it might be unclear how these properties can be used to obtain CR-expressions which evaluate more efficiently. However, by considering polynomial CRs, we get:

Lemma 4. Let $\{\varphi_0, +, \varphi_1, +, \ldots, +, \varphi_k\}$ be a polynomial CR with $\varphi_0, \ldots, \varphi_k \in \boldsymbol{P}$. Then,

$$\begin{aligned}\sin(\{\varphi_0, +, \ldots, +, \varphi_k\}) &= \Im(\{e^{i\varphi_0}, *, e^{i\varphi_1}, *, \ldots, *, e^{i\varphi_k}\}) \\ &= \Im(\{\cos\varphi_0 + i\sin\varphi_0, *, \ldots, *, \cos\varphi_k + i\sin\varphi_k\}),\end{aligned} \tag{15}$$

$$\begin{aligned}\cos(\{\varphi_0, +, \ldots, +, \varphi_k\}) &= \Re(\{e^{i\varphi_0}, *, e^{i\varphi_1}, *, \ldots, *, e^{i\varphi_k}\}) \\ &= \Re(\{\cos\varphi_0 + i\sin\varphi_0, *, \ldots, *, \cos\varphi_k + i\sin\varphi_k\}).\end{aligned} \tag{16}$$

Hence, for some common cases, Lemma 4 enables us to replace the computationally rather expensive trigonometric operations by multiplications of complex numbers.

It is also interesting to observe that, with regard to the primitive arithmetic operations required for their evaluation, the here considered complex CRs are closely related to the sin/cos-type SRRs suggested in Zima (1992). Those sin/cos-type SRRs do not use complex numbers to describe the evaluation of trigonometric expression but use two interwoven SRRs which interact in a fashion similar to the multiplication of complex numbers. As one of the referees pointed out, this scheme defines all computations explicitly in terms of operations from the same domain and hence also works for polynomial CRs with complex CR-components. However, the definition of sin/cos-type SRRs leads to rather cumbersome new notations and requires to reconsider and restate most of the simplification rules as well as evaluation procedures for SRRs of this type.

Recalling that CR-expressions are functions defined over **N**, there are of course further simplification properties which do not only apply to CR-expressions, but to functions in general. Assuming that f, g are functions where the range of g is not complex, there are, for example, the following simplification rules:

$$0 + f = 1 * f = ff, \qquad 1 * f = f; \tag{17}$$

$$g * \Re(f) = \Re(g * f), \qquad g * \Im(f) = \Im(g * f). \tag{18}$$

2.5 The algorithm CRMake

Let us summarize this section by outlining the algorithm `CRMake` which, for a given closed-form function $G(x)$, constructs an equivalent and simplified CR-expression Φ. Again, see Bachmann et al. (1994) for more details. Loosely speaking, `CRMake`(G, x, x_0, h) works as follows:

1. the input expression G is converted into an equivalent CR-expression Φ by replacing expressions of the form x^j by the polynomial CR $\{x_{0j}, +, x_{1j}, +, \ldots, +, x_{j-1,j}, +, x_{jj}\}$ where the constants x_{pq} are obtained by (12) with $x_{01} = x_0$, $x_{11} = h$.
2. Φ is recursively simplified by the simplification rules (3)–(18).

Three further remarks about the algorithm: First, the cost index of the CR-expression returned by `CRMake` is minimized with respect to the results of Sect. 2.4. Second, the polynomial CR returned by `CRMake` for a polynomial as input is in normal form, viz., if $p(x)$ and $q(x)$ are polynomials with $p(x) = q(x)$ for all x, then `CRMake`(p, x, x_0, h) is syntactically equivalent to `CRMake`(q, x, x_0, h). Third, if G does not contain any further indeterminates then the constant-coefficients appearing in the returned CR-expressions are of the same type as the input parameters x_0 and h. Otherwise, symbolic CR-expressions are returned which can be initialized and "reused" for different values of the contained indeterminates (the same applies if the values of x_0 and h are left undetermined).

Let us illustrate the functionality of the algorithm `CRMake` by an example. Consider $G(x) = \sin(x)\, e^{x^2}$ and let $x_0 = 0,\ h = 0.1$. As a result of step 1, G is first transformed into the equivalent CR-expression

$$\Phi = \sin(\{x_0, +, h\})\, e^{\{x_0^2, +, 2hx_0, +, h^2, +, 2h^2\}} = \sin(\{0, +, 0.1\})\, e^{\{0, +, 0.01, +, 0.02\}}.$$

Next, Φ is recursively simplified

$$\begin{aligned} \sin(\{0, +, 0.1\}) &\Longrightarrow \Im(\{e^0, *, e^{0.1i}\}), \\ e^{\{0, +, 0.01, +, 0.02\}} &\Longrightarrow \{1, *, e^{0.01}, *, e^{0.02}\}, \\ \Im(\{1, *, e^{0.1i}\})\{1, *, e^{0.01}, *, e^{0.02}\} &\Longrightarrow \Im(\{1, *, e^{0.01+0.1i}, *, e^{0.02}\}) \end{aligned}$$

by (15), (8), (6), and (18) (in that order), which finally results in $\Phi = \Im(\{1, *, \cos(0.01) + i \sin(0.1), *, e^{0.02}\})$. Hence, only 8 primitive arithmetic operations (4 multiplications and 2 additions for multiplying two complex numbers and 2 multiplications for multiplying a complex number by a real number) are required to obtain the value of Φ at one point.

3 Chains of recurrences for functions of two variables

Given a function $G(x, y)$ of two variables which we wish to evaluate for $G(x_0 + nh_x, y_0 + mh_y)$ for $n = 0, \ldots, N,\ m = 0, \ldots, M$, we examine in this section how the one-dimensional CR method can be extended to such two-dimensional evaluations.

3.1 Two-dimensional CR-expressions

First, let us illustrate the main idea behind two-dimensional CR-expressions by an example. Consider the function $G(x, y) = x + y$ for which CRMake(G, x, x_0, h_x) returns the CR $\Phi_x = \{x_0 + y + h_x\}$. By again calling CRMake with the CR-components of Φ_x and y, y_0, h_y as arguments, we get the following:

$$\texttt{CRMake}(x_0 + y, y, y_0, h) = \{x_0 + y_0, +, h_y\}, \quad \texttt{CRMake}(h_x, y, y_0, h_y) = h_x.$$

Putting everything together we obtain the CR $\Phi_{xy} = \{\{x_0 + y_0, +, h_y\}, +, h_x\}$ which contains a CR in y-direction as a component. Hence, instead of initializing Φ_x for each value of y we simply evaluate the inner CR of Φ_{xy} which yields a "ready-to-evaluate" CR in x-direction.

However, there is an ambiguity problem with this approach. Consider the function $G(x, y) = y + e^x$ and let $x_0 = y_0 = h_x = h_y = 2$. By applying the technique described above, we obtain the CR-expression $\Phi_{xy} = \{2, +, 2\} + \{e^2, *, e^2\}$ for G. Notice that there is no unique way to determine the correct evaluation procedure of this representation. Quite different results (and possibly quite wrong ones) will be obtained for different interpretations of the x- and/or y-direction of the CRs contained in Φ_{xy}. Hence, we have to "tag" the CRs with their "evaluation direction" in order to obtain a unique and correct evaluation procedure of two-dimensional CR-expressions.

More precisely, we say that Φ_{xy} is a *two-dimensional* CR-*expression* if it represents one of the following functions over $\mathbf{N} \times \mathbf{N}$.

1. A one-dimensional CR-expression Φ_y as defined in Sect. 2.3 for which we define $\Phi_{xy}(n, m) = \Phi_y(m)$ for all n.
2. A CR of the form $\{\Phi_{0y}, \odot_1, \Phi_{1y}, \odot_2, \ldots, \Phi_{k-1,y}, \odot_k, f_k\}_x$ where the CR-components $\Phi_{0y}, \Phi_{1y}, \ldots \Phi_{k-1,y}$ are one-dimensional CR-expressions as defined in Sect. 2.3 and $f_k(i, j)$ is a two-dimensional CR-expression. In this case we define $\Phi_{xy}(n, m) = \{\Phi_{0y}(m), \odot_1, \Phi_{1y}(m), \ldots, \odot_k, f_k(i, m)\}_x(n)$.
3. $F(\Phi_1, \Phi_2, \ldots, \Phi_l)$, where F is a function of m arguments and $\Phi_1, \Phi_2, \ldots, \Phi_l$ are two-dimensional CR-expressions.

Loosely speaking, the main difference between a one-dimensional and a two-dimensional CR-expression is that a two-dimensional CR-expression may contain "tagged" or two-dimensional CRs of the form $\{\Phi_{0y}, \odot_1, \ldots, \Phi_{k-1,y}, \odot_k, f_k\}_x$ in which case the CR-components may not only be constants, but also one-dimensional CR-expressions. Furthermore, it is clear from this notation which CRs correspond to which evaluation direction. Let's look at some examples of functions $G(x, y)$ and equivalent, simplified CR-expressions Φ_{xy} (by equivalent we mean here $G(x_0+n\,h_x, y_0+m\,h_y) = \Phi_{xy}(n,m)$ for all $n, m \geq 0$):

$$
\begin{aligned}
x &= \{x_0, +, h_x\}_x; \quad y = \{y_0, +, h_y\}_y;\\
x+y &= \{\{x_0+y_0, +, h_y\}_y, +, h_x\}_x;\\
x*y &= \{\{x_0\,y_0, +, x_0 h_y\}_y, +, \{h_x\,y_0, +, h_x\,h_y\}_y\,\}_x;\\
y+e^x &= \{y_0, +, h_y\}_y + \{e^{x_0}, *, e^{h_x}\}_x;\\
x^2\,y &= \{\{x_0^2 y_0, +, x_0^2 h_y\}_y, +, \{h_x y_0(2x_0+h_x), +, h_x h_y(2x_0+h_x)\}_y,\\
&\qquad +, \{2h_x^2 y_0, +, 2h_x^2 h_y\}_y\}_x;\\
\cos(xy) &= \Re(\{\{e^{i\,x_0 y_0}, *, e^{i\,x_0 h_y}\}_y, *, \{e^{i\,h_x y_0}, *, e^{i\,h_x h_y}\}_y\}_x);\\
e^x \sin y &= \{\,\Im(\{e^{x_0+i\,y_0}, *, e^{x_0+i\,h_y}\}_y), *, e^{h_x}\}_x;\\
e^y \cos x &= \Re(\{\{e^{y_0+i\,x_0}, *, e^{h_y+i\,x_0}\}_y, *, e^{i\,h_x}\}_x).
\end{aligned}
$$

3.2 Construction of two-dimensional CR-expressions

Two-dimensional CR-expressions are constructed just like one-dimensional CR-expressions, namely, by first transforming a given expression $G(x, y)$ into an equivalent two-dimensional CR-expression Φ_{xy}, followed by a simplification of Φ_{xy}. First, let us consider the algorithm `CRConvert` which converts of a given expression in two indeterminates into an equivalent two-dimensional CR-expression.

CRConvert $(G, x, x_0, h_x, y, y_0, h_y)$. For a given closed-form input expression $G(x, y)$, this algorithm returns an equivalent two-dimensional CR-expression Φ_{xy} such that $\Phi_{xy}(n, m) = G(x_0 + n\,h_x, y_0 + m\,h_y)$ for all $n, m > 0$.

C1 [Base cases]
If G is a constant then return G.
If $G \equiv x$ return $\{x_0, +, h_x\}_x$.
If $G \equiv y$ return $\{y_0, +, h_y\}_y$.

C2 [Convert integral powers of the variables]
If $G \equiv x^j$, $j \in \mathbf{N}$ then return $\{x_{0j}, +, x_{1j}, +, \ldots, +, x_{j-1,j}, +, x_{jj}\}_x$ where the constants x_{pq} are obtained by (12) with $x_{01} = x_0, x_{11} = h_x$. If $G \equiv y^j$, $j \in \mathbf{N}$ then return $\{y_{0j}, +, y_{1j}, +, \ldots, +, y_{j-1,j}, +, y_{jj}\}_y$ where the constants y_{pq} are obtained by (12) with $y_{01} = y_0, y_{11} = h_y$.

C3 [Recursive step]
Apply the algorithm recursively to all arguments of G. Replace the original arguments of G with their returned two-dimensional CR-expressions and return the result of this substitution.

Next, we discuss the simplification of two-dimensional CR-expressions starting with the important observation that *all simplification rules of Sect. 2.4* do not only apply to CRs of the form $\{\varphi_0, \odot_1, \varphi_1, \ldots, \odot_k, f_k\}$ (i.e., to one-dimensional CRs), but also to CRs of the form $\{\Phi_{0y}, \odot_1, \Phi_{1y}, \ldots, \odot_k, f_k\}_x$ (i.e., to two-dimensional CRs).

Consequently, all we need to reconsider are simplification rules involving operations of two-dimensional CRs with one-dimensional CR-expressions.

Lemma 5. Let $\{\Phi_{0y}, \odot_1, \Phi_{1y}, \odot_2, \ldots, \Phi_{k-1,y}, \odot_k, f_k\}_x$ be a two-dimensional CR and Φ_y be a one-dimensional CR-expression. Then,

$$\Phi_y + \{\Phi_{0y}, +, \Phi_{1y}, \odot_2, \ldots, f_k\}_x = \{\Phi_y + \Phi_{0y}, +, \Phi_{1y}, \odot_2, \ldots, f_k\}_x; \tag{19}$$

$$\Phi_y * \{\Phi_{0y}, *, \Phi_{1y}, \odot_2, \ldots, f_k\}_x = \{\Phi_y * \Phi_{0y}, *, \Phi_{1y}, \odot_2, \ldots, f_k\}_x; \tag{20}$$

$$\Phi_y * \{\Phi_{0y}, +, \Phi_{1y}, \odot_2, \ldots, f_k\}_x = \{\Phi_y * \Phi_{0y}, +, \Phi_y * \{\Phi_{1y}, \odot_2, \ldots, f_k\}_x\}_x; \tag{21}$$

$$\Phi_y^{\{\Phi_{0y}, +, \Phi_{1y}, \odot_2, \ldots, f_k\}_x} = \{\Phi_y^{\Phi_{0y}}, *, \Phi_y^{\{\Phi_{1y}, \odot_2, \ldots, f_k\}_x}\}_x; \tag{22}$$

$$\{\Phi_{0y}, *, \Phi_{1y}, \odot_2, \ldots, f_k\}_x^{\Phi_y} = \{\Phi_{0y}^{\Phi_y}, *, \{\Phi_{1y}, \odot_2, \ldots, f_k\}_x^{\Phi_y}\}_x. \tag{23}$$

3.3 The algorithm CRMake2

Let us summarize this section by describing the algorithm `CRMake2`.

CRMake2 $(G, x, x_0, h_x, y, y_0, h_y)$. For a given closed-form input expression $G(x, y)$, this algorithm returns a simplified, two-dimensional CR-expression Φ_{xy} such that $\Phi_{xy}(n, m) = G(x_0 + n\,h_x, y_0 + m\,h_y)$ for all $n, m \geq 0$.

S1 [Conversion]
Set $\Phi_{xy} \longleftarrow$ `CRConvert`$(G, x, x_0, h_x, y, y_0, h_y)$.

S2 [Simplification]
Recursively simplify Φ_{xy} (starting from the leaves of the expression-tree of Φ_{xy}) by the simplification rules (3)–(23) and return the simplified two-dimensional CR-expression Φ_{xy}.

Informally speaking, the algorithm does mainly the following:

1. Polynomials are expanded and replaced by (possibly two-dimensional) polynomial CRs.
2. Products of exponentials which have a constant base and a polynomial CR as exponents are replaced by exponential CRs.
3. Trigonometric functions of polynomial CRs are replaced by exponential CRs with complex coefficients.
4. If possible, some further simplifications are applied (logarithms and exponentials of exponential CRs, factorials of polynomial CRs, general simplification, etc.).

Let us illustrate the way the algorithm works by considering some transformation and simplification steps of `CRMake2`$(\cos(x\,y), x, x_0, h_x, y, y_0, h_y)$:

$$\cos(x\,y) \Rightarrow \cos(\{x_0, +, h_x\}_x * \{y_0, +, h_y\}_y) \text{ after CRConvert},$$

$$\Rightarrow \cos(\{x_0 * \{y_0, +, h_y\}_y, +, h_x * \{y_0, +, h_y\}_y\}_x) \text{ by (20)},$$

$$\Rightarrow \cos(\{\{x_0\,y_0, +, x_0 h_y\}_y, +, \{h_x\,y_0, +, h_x\,h_y\}_y\}_x) \text{ by (7)},$$

$$\Rightarrow \Re(\{e^{i\,\{x_0\,y_0, +, x_0 h_y\}_y}, *, e^{i\,\{h_x\,y_0 + h_x\,h_y\}_y}\}_x) \text{ by (16)},$$

$$\Rightarrow \Re(\{\{e^{i\,x_0 y_0}, *, e^{i\,x_0 h_y}\}_y, *, \{e^{i\,h_x y_0}, *, e^{i\,h_x h_y}\}_y\}_x) \text{ by (8)}.$$

4 A Maxima/IZIC implementation

We implemented the algorithms described above in the Common Lisp (CL) level of Maxima (The MATHLAB Group 1977) [AKCL-based Maxima version 4.15; it also runs under the Macsyma version 418 (Macsyma 1993)] and applied it to the graphing system IZIC (Fourier et al. 1993). When implementing the CR method and applying it to surface computations one has to answer several questions: Which data structures should be used to represent CR-expressions? How can generality and efficiency be optimally balanced? How should the implementation be applied to a graphing system? How do timings of the implementation compare with other evaluation methods? These and several other topics are discussed in this section.

4.1 The Maxima part

The implementation consists of approximately 4000 lines of AKCL code. It is based on our preliminary implementation of the CR technique reported in Bachmann et al. (1994). However, a completely new implementation was necessary because we needed a new internal Lisp representaion of two-dimensional CRs. The implementation consists of the following three major parts which implement the above described algorithms for constructing and evaluating CR-expressions.

CR conversion. The Lisp-representation of a Maxima expression is converted into an internal Lisp-representation of CR-expressions as described by the algorithm `CRConvert`. A CR-expressions is internally represented by a list of the following form: `(Header arg1 arg2 ... (argk).Header` is a list containing specific information about how to interpret the remaining arguments (e.g., as a polynomial CR in x-direction, as an addition of CR-expressions, etc.). The remaining arguments `arg1`, `arg2,...,argk` might be constants or again CR-expressions. Hence, we even represent CRs internally just like "normal" expressions, which turned out to be very convenient and efficient. Furthermore, the coefficients x_{pq} of the polynomial CRs corresponding to x^j (step C2 of the algorithm `CRConvert`) are stored in a global array in order to avoid unnecessary recomputations.

CR simplification. A given CR-expression is recursively simplified by unconditionally applying the simplification rules (3)–(23).

CR evaluation. Various routines for the evaluation of one and/or two dimensional CRs and CR-expressions were implemented. See below for more details.

The implementation was designed to construct and evaluate CRs efficiently for any valid Maxima domain (rationals, floating-point numbers, bfloat numbers, symbolic expressions). All necessary arithmetic operations were implemented by macros. Hence, if the computational domain is known for a specific application (like floating-point numbers for plotting), those macros can be changed (to declared `long-float` Lisp operations, for example) in order to obtain more efficient code.

4.2 IZIC

IZIC is a stand-alone 3-D graphic tool allowing visualization of mathematical objects such as curves and surfaces (Fournier et al. 1993). In short, IZIC capabilities include management of the illumination model, shading, transparency, etc. IZIC consists of a hardware-independent C library which provides a large collection of graphical operations and which is linked to a TCL interpreter. At runtime, IZIC can be driven through its graphic control panel and/or via TCL scripts which are sent remotely from one or more computer algebra systems (CAS). Currently, there are four interfaces to IZIC, from Maple, Mathematica, Reduce, and Maxima. We implemented the Maxima interface (called MaxIZIC) in order to demonstrate and test the application of the CR technique to plotting. For more details about MaxIZIC, see Bachmann (1994).

Graphic objects to be visualized by IZIC have to be defined in the ZIC format which is communicated using files. Files in this format include a series of optional and mandatory fields: type of the objects, bounding box, colors, object encoding, etc. Typically, *object encoding* is a long collection of floats whose structure depends on the type of the object.

Hence, the role of an interface between a CAS and IZIC is to generate and output representations of curves and surfaces in the ZIC format and to communicate with IZIC by sending TCL scripts in order to remotely display and manipulate objects.

Typically such an interface provides a collection of user-level commands (such as `izplot`, `izplot3D`, `izsurface`, `izsphere`, etc.) which extend or supplement existing graphing capabilities of the CAS. However, as we pointed out in the introduction, the naive and unoptimized function evaluation procedures used by these interfaces often result in one of their major bottlenecks, namely inefficient and time-consuming computations. As we will show in the next section, CRs offer a remedy.

4.3 Putting CRs and IZIC together

Maxima top-level functions which generate objects to be displayed by IZIC generally have the following calling sequence:

```
izfname(expressions, variables, {opt-specs})
```

where `izfname` is the name of a function (e.g., `izplot3D`, `izcurve`, `izsurface`), `expressions` is a list of mathematical expressions (e.g., `[fx, fy, fz]`) specifying the shape of the object, `variables` is a list specifying the evaluation variables of `expressions` and their evaluation ranges, and `opt-specs` includes optional specification of the color, grid, transparency, etc. The object-generating functions first evaluate `fx`, `fy`, and `fz` for all points, then write the needed geometrical and numerical specifications in a file using the ZIC format, and finally call IZIC to display the content of the file.

The evaluation routine used by the object-generating functions is determined at runtime by the value of the global variable `izev`. To test our CR programs, we implemented several evaluation routines:

- `nev`- normal (or naive), unoptimized Maxima evaluations of a given expression. Translated into the Maxima top-level programming language, the main loop for the evaluation of one-dimensional expressions looks as follows:

```
numer:true;  xi:x0;
for i:0 thru n-1 do
  (Result[i]:ev(f,x=xi);  xi:xi+h;);
```

- `fnev` – Lisp-level floating-point evaluations of a "Horner-simplified" expression as suggested by Wang (1990). This is done by converting the expressions defined in Maxima representation into a corresponding Lisp representation and by simultaneously applying "Horner's rule" as much as possible. For example, the Maxima expression `x^2+x+1` is internally represented by Maxima as

```
((MPLUS SIMP) 1 $X ((MEXPT) $X 2))
```

  and converted to the Lisp expression

```
(+ 1 (* *x* (+ 1 *x*)))
```

  which then can be evaluated by the following loop:

```
;; lisp-f is a lisp-expression containing *x* as variable
(setq *x* x0)
(dotimes (i n)
        (setf (aref result i) (eval lisp-f))
        (setq *x* (+ *x* h)))
```

- `crev` – first, for a given Maxima expression an equivalent CR is constructed which is then evaluated. Both the CR construction and the CR evaluation are done using Maxima operations. For a one-dimensional expression, some fragments of the source code are as follows:

```
(defmacro myplus (op1 op2)
   '(meval '((MPLUS) ,op1 ,op2)))
...
(setq fcr (crmake $f x x0 h))
...
(if (polynomial-cr-p fcr)
  ;; now fcr is of the form (header f0 f1 f2 .... fl)
  (do* ((crlength (1- (list-length (cdr fcr))))
       (cra (make-array (1+ crlength)
                        :initial-contents (cdr fcr)))
       (i 0 (1+ i)))
      ((= i n))
     (setf (aref result i) (aref cra 0))
     (dotimes (j crlength)
           (setf (aref cra j)
                 (myplus (aref cra j)
                         (aref cra (1+ j)))))))
```

- `fcrev` – similar to `crev`, except that the CR construction and CR evaluation is done using floating-point Lisp operations and for the CR evaluation routines attention has been paid to "fast floating-point processing in Common Lisp" as advocated in Fateman et al. (1995). For example, macros like the following are used to define floating-point arithmetic operations:

```
(defmacro myplus (op1 op2)
   '(the long-float (+ (the long-float ,op1)
                       (the long-float ,op2))))
```

Notice that these code fragments are only simplified renditions of the routines we actually implemented. Some further technical optimizations, like loop-unrolling and type-declarations, were applied where appropriate. However, the code fragments illustrate the basic techniques we used in the respective evaluation routines.

4.4 Results and timings

Table 1 shows some timings of our programs. The drawings generated by IZIC together with more detailed surface specifications and a script of the Maxima session are given in the appendix.

All timings are in seconds and were obtained on a lightly loaded Sun SPARCstation2 with 32 MByte main memory. The timings do neither include writing out the data into the ZIC file nor display by IZIC.

We can observe that CRs can be constructed and evaluated very efficiently – the speedups gained by an application of the two-dimensional CR technique are tremendous and considerably higher than those of an earlier implementation, reported in Bachmann et al. (1994). For polynomial and exponential CRs, we gain speedups in the range of 100 – our earlier implementation gained speedups in the range of 3–20. Furthermore, it is very much worth the effort to replace Maxima expressions by Lisp expressions (compare `nev` with `fnev` and `crev` with `fcrev`) and to use especially optimized routines for floating-point computations (compare `crev` with `fcrev`).

It might also be interesting to observe that the timings for example 5 reported by Wang (1990) are 121 s for Vaxima evaluation (equivalent to `nev`) and 23 s for Lisp evaluation (equivalent to `fnev`). He obtained these timings on an Encore multiprocessor machine with 12 CPU each rated at 0.75 MIPS and made the remark: "The lisp level evaluation is not exactly fast but is tolerable, the Vaxima level evaluation is very slow." With our current machines and technology at hand, we can rephrase this remark as:

The Lisp-level CR evaluation is sufficiently fast, the Maxima-level CR evaluation is not exactly fast, but tolerable.

The competitive advantage of our CR implementation can be traced back to the following reasons: fewer arithmetic operations have to be performed in comparison to `fnev`; numerically rather expensive operations (like sin and exp) are replaced by additions and multiplications; a high degree of "precompilation" can be achieved because

Table 1. Timings of different evaluation routines in seconds

	Surface	Grid	nev	fnev	crev	fcrev
1	$\left[\frac{u}{2}-\frac{u^3}{6}+\frac{uv^2}{2},\ \frac{v^3}{6}-\frac{v}{2}-\frac{vu^2}{2},\ \frac{u^2}{2}-\frac{v^2}{2}\right]$	50×50	60.13	9.37	6.93	0.10
2	$[\cos(x\,y),\ x,\ y]$	50×50	35.95	1.23	3.00	0.05
3	$[e^x\sin(y),\ x,\ y]$	50×50	41.12	2.38	0.95	0.05
4	$[r\cos(t),\ r\sin(t),\ t]$	15×45	10.28	0.55	1.23	0.03
5	$[(1-r^2)\cos(s),\ (1-r^2)\sin(s),\ r]$	30×30	17.07	2.10	1.92	0.05

of the simple structure of polynomial and exponential CRs; by careful implementations of floating-point routines in Lisp (as done for the evaluation routines of `fcrev`), one can almost achieve the efficiency of equivalent C or Fortran programs.[2] The latter is one of the major conclusions of Fateman et al. (1995) and our results very much confirm this. For example, the higher speedups (for polynomial and exponential CRs) achieved by our current implementation in comparison with those reported in Bachmann et al. (1994) are mainly due to that reason.

One further question should be addressed here, namely that of the numeric stability of CR evaluations. Although algebraically exact, CR evaluations are vulnerable to numeric instabilities. This is mainly due to the fact that numeric errors can propagate and accumulate because intermediate results are extensively reused. Hence, if the numeric error exceeds a certain limit, we might have to "refresh" the CR evaluation by reconstructing the constants appearing in CRs. However, our experiments indicate that such means are not necessary for surface computations (the maximal relative numeric error we encountered was about 10^{-9}): On the one hand, relatively few successive evaluations which can accumulate errors are performed (e.g., on the 50×50 grid the error can accumulate over a maximum of 100 evaluations) and, on the other hand, the accuracy needed is not very high (say, not more than 5 accurate digits). Nevertheless, more work should be done in order to reliably control the numerical accuracy of CR evaluations.

5 Conclusions and future work

The main results of this paper are the following.

We extended the theoretical concepts of CRs in order to expedite the evaluation of trigonometric functions and functions of two variables and gave algorithms for the construction, simplification, and evaluation of two-dimensional CRs.

We implemented the described algorithms in Maxima/Common Lisp. Our implementation can be used to make evaluations of closed-form functions over linearly sampled points more efficient. The implementation works over any valid Maxima domain, but is especially effective for floating-point evaluations.

We interfaced our Maxima implementation with the IZIC graphing tool and successfully demonstrated the practical feasibility of our method to make curve and surface computations reasonably fast.

The CR method guarantees more efficient evaluations for expressions containing polynomial or trigonometric subexpressions, can be applied to any valid, closed-form expression (which might even contain user-defined operators) and is in the worst case just as expensive as straightforward evaluations of the expressions considered.

However, much future work lies ahead about theoretical and algorithmic aspects of the CR methods as well as about its implementation and application to plotting. First, the theoretical concepts of CRs should be extended to functions of more than two variables and to other classes of primitive functions like the logarithmic and hyperbolic

---

2 Some further experiments we conducted showed that even compiled C-programs for "Horner evaluations" of a polynomial are still slower than our Lisp CR-routines and that compiled C-programs for evaluations of polynomial CRs (i.e., of polynomials) are only marginally faster than our Lisp CR-routines.

functions and the inverse trigonometric and inverse hyperbolic functions. Second, from an algorithmic point of view, we should consider the construction of optimal CRs. Among others, the following problems should be addressed.

- What is the best ordering of the evaluation variables?
  Consider a polynomial $p(x, y)$. With our current technique, we construct an equivalent polynomial CR Φ_{xy} of the form $\{\{\ldots\}_y + \ldots + \{\ldots\}_y\}_x$. However, with little change we could also construct an equivalent CR Φ_{yx} which then would look like $\{\{\ldots\}_x + \ldots + \{\ldots\}_x\}_y$. In other words, we can change the ordering of the variables w.l.o.g. which could result in considerably faster evaluations. Take $x^5 y^2$, for example, and let $n_x = 10$, $n_y = 10$. To evaluate Φ_{xy} for those 100 points requires 600 additions, but an evaluation of Φ_{yx} requires only 300 additions.
- How can we efficiently find "common sub-CRs and sub-CR-expressions"?
- How can we set up a scheme of effective conditional simplifications?
  With our current method, simplification rules of CR-expressions are always unconditionally applied. However, there are cases where we could obtain CR-expressions which evaluate more efficiently by not applying certain rules at certain times.

We believe that these problems do not have a solution which is efficient *and* general. More information about the time characteristics of the underlying computational domain is needed and backtracking CR construction algorithms seem unavoidable. However, can we find efficient and effective heuristic approaches to solve those problems?

Because of the ability to write Maxima programs in Lisp, Maxima has the advantage over most other CAS that, if necessary, programs can be written on a very low level with control over almost all ongoing computations. This might sometimes be very tedious, but, as we discussed earlier, the success of our current implementation very much depends on this low-level control. Naturally the question arises how and how successfully the CR method can be applied in other CAS and "mathematical engines" where the situation concerning low-level programming might be different.

The here presented application of the CR method to plotting is limited to *explicitly* defined curves and surfaces which can be computed over a rectangular domain. Extensions to implicitly defined surfaces (i.e., surfaces defined by equations of the form $f(x, y, z) = 0$), to surfaces defined over circular domains and to surfaces which are defined by various parameterizations or nested expressions should be considered. However, it should be noted that the more piecewise a surface is defined, the less successfully the CR method can be applied: a certain number of evaluations are required for the CR methods to be effective.

As a summary, we think that the CR technique is an important key to fast and efficient plotting and that further investigations could increase the efficiency, generality, and portability of this technology.

6 Availability

IZIC is freely available via anonymous ftp from zenon.inria.fr:/safir/. The archive file includes the Maple, Mathematica, Reduce, and Maxima interfaces. The CR extension of the Maxima interface, described in Sect. 4, can be obtained from the author.

Appendix: a Maxima session and IZIC drawings

```
(C1) load("maxizic.o");
Loading maxizic.o
[compiled in AKCL 1-615] start address -T 6bb000
(D1)                                    #maxizic.o
(C2) load("fcr.o");
Loading fcr.o
[compiled in AKCL 1-615] start address -T 6c1000
(D2)                                    #fcr.o
(C3) izplot3D([x,cos(x*y),y],[[x,-3,3],[y,-3,3]],[50,50]);
fcrev time:
real time : 0.067 secs
run time  : 0.050 secs
(D3)                                    DONE
(C4) (izev:fnev,
      izplot3D([x,cos(x*y),y],[[x,-3,3],[y,-3,3]],[50,50]));
fnev time:
real time : 1.250 secs
run time  : 1.233 secs
(D4)                                    DONE
(C5) (izev:fcrev,
      izplot3D([x,y,sin(x)*exp(y)],[[x,-5,5],[y,-2,2]]));
fcrev time:
real time : 0.100 secs
run time  : 0.050 secs
(D5)                                    DONE
(C6) (izev:nev,
      izplot3D([x,y,sin(x)*exp(y)],[[x,-5,5],[y,-2,2]]));
```

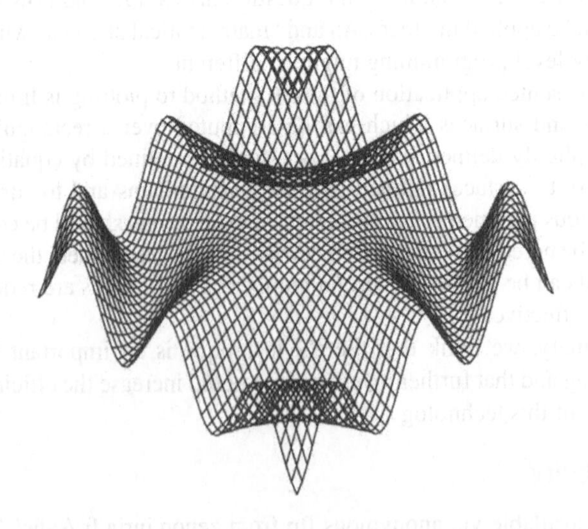

Fig. 1. $\cos(x\,y)$ rotated by $\alpha = 45°$, $\beta = 45°$

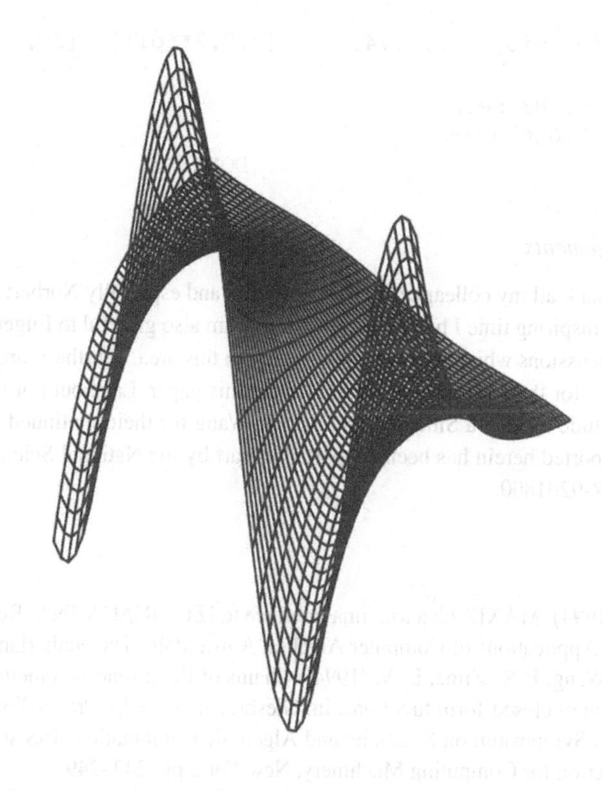

Fig. 2. $e^x \sin(y)$ rotated by $\alpha = 132°$, $\gamma = 47°$

```
nev time:
real time : 43.833 secs
run time  : 41.117 secs
(D6)                          DONE
(C7) ship:[(1-r^2)*cos(s),(1-r^2)*sin(s),r];

                               2                 2
(D7)                  [(1 - R ) COS(S),(1 - R )SIN(S), R]]
```

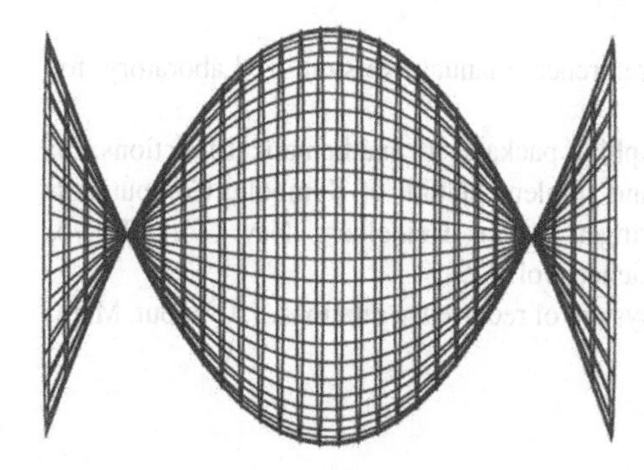

Fig. 3. P. S. Wang's "ship" example rotated by $\beta = 90°$

```
(C8) izplot3D(ship, [[r,-1.4,1.4],[s,0,2*%pi]], [30,30]);
fcrev time:
real time : 0.083 secs
run time  : 0.050 secs
(D8)                                DONE
```

Acknowledgements

I would like to thank all my colleagues at RIACA/CAN, and especially Norbert Kajler, for the very fruitful and inspiring time I had at their institute. I am also grateful to Eugene V. Zima for the profitable discussions which fostered my research in this area. I furthermore would like to thank the referees for their valuable comments about this paper. Last, but not least, I wish to express my gratitude to Meera Sitharam and Paul S. Wang for their continued support of my studies. Work reported herein has been supported in part by the National Science Foundation under Grant CCR-9201800.

References

Bachmann, O. (1994): MAXIZIC: a Maxima interface to IZIC. RIACA Tech. Rep. 4, Research Institute for Applications of Computer Algebra, Amsterdam, The Netherlands.

Bachmann, O., Wang, P. S., Zima, E. V. (1994): Chains of Recurrences: a method to expedite the evaluation of closed-form functions. In: Giesbrecht, M. (ed.): Proceedings of the ACM International Symposium on Symbolic and Algebraic Computation (ISSAC '94), Oxford, UK. Association for Computing Machinery, New York, pp. 242–249.

Char, B. W., Geddes, K. O., Gonnet, G. H., Monagan, M. B., Watt, S. M. (1988): Maple reference manual, 5th edn. Watcom Publications, Waterloo, Ont.

Fateman, R. J., Broughan, K. A., Willcock, D. K., Rettig, D. (1995): Fast floating-point processing in Common Lisp. ACM Trans. Math. Software 21: 26–62.

Fournier, R., Kajler, N., Mourrain, B. (1993): IZIC: a portable language-driven tool for mathematical surfaces visualization. In: Miola, A. (ed.): Proceedings of Design and Implementation of Symbolic Computation Systems (DISCO '93), Gmunden, Austria. Springer, Berlin Heidelberg New York Tokyo, pp. 341–353 (Lecture notes in computer science, vol. 722).

Kaltofen, E. (1988): Greatest common divisors of polynomials given by straight-line programs. J. ACM 35: 231–264.

Knuth, D. E. (1981): The art of computer programming, vol. 2, seminumerical algorithms. Addison-Wesley, Reading, MA.

Macsyma (1993): Macsyma mathematics reference manual, version 14. Macsyma Inc., Arlington, MA.

The MATHLAB Group (1977): Macsyma reference manual, version 9. Laboratory for Computer Science, MIT, Cambridge, MA.

Wang, P. S. (1990): A system independent graphing package for mathematical functions. In: Miola, A. (ed.): Proceedings of Design and Implementation of Symbolic Computation Systems (DISCO '90), Capri, Italy. Springer, Berlin Heidelberg New York Tokyo, pp. 245–254 (Lecture notes in computer science, vol. 429).

Zima, E. V. (1984): Automatic construction of system of recurrence relations. J. Comput. Math. Math. Phys. 24: 193–197.

Zima, E. V. (1992): Recurrent relations and speed-up of computations using computer algebra systems. In: Fitch, J. (ed.): Proceedings of Design and Implementation of Symbolic Computation Systems (DISCO '92), Bath, UK. Springer, Berlin Heidelberg New York Tokyo, pp. 152–161 (Lecture notes in computer science, vol. 721).

Design principles of Mathpert: software to support education in algebra and calculus

Michael Beeson

1 Introduction

This paper lists eight design criteria that must be met if we are to provide successful computer support for education in algebra, trigonometry, and calculus. It also describes Mathpert, a piece of software that was built with these criteria in mind. The description given here is intended for designers of other software, for designers of new teaching materials and curricula utilizing mathematical software, and for professors interested in using such software. The design principles in question involve both the user interface and the internal operation of the software. For example, three important principles are *cognitive fidelity*, the *glass box* principle, and the *correctness* principle. After an overview of design principles, we discuss the design of Mathpert in the light of these principles, showing how the main lines of the design were determined by these principles. (The scope of this paper is strictly limited to an exposition of the design principles and their application to Mathpert. I shall not attempt to review projects other than Mathpert in the light of these design principles.)

2 Purposes of software for mathematics education

The first step in proper software design is a clear statement of the purpose of the software. Here, for example, is a statement of purpose for Mathpert:

Mathpert is intended to replace paper-and-pencil homework in algebra, trigonometry, and calculus, retaining compatibility with the existing curriculum while at the same time supporting innovative curriculum changes; to provide easy-to-use computer graphics for classroom demonstrations in those subjects, as well as for home study; to replace or supplement chalk-and-blackboard in the classroom for symbolic problems as well as graphs. Mathpert is not intended to replace teachers or books: it is not explicitly tutorial. It is a "computerized environment for solving problems." It can be used by students of all ages and levels who are prepared to learn the subject matter.

Most experiments to date with using software in calculus instruction have used general-purpose symbolic programs such as Maple or Mathematica. These programs were not developed specifically for education, and it is therefore small wonder that their capabilities are not entirely matched to the needs of students.

by John Stasko. While the framework contains extensible models for images and transitions, various types of images and transitions are provided within an initial library of functions. These primitives give a compact working set for easy learning, yet they still provide sufficient depth to generate moderately complex animation sequences. More recently, the Zeus system uses sound and 3-D interactive graphics to reach higher levels of expressiveness (Brown 1992, Brown and Hershberger 1991, Brown and Najork 1993). Zeus is a system for the animation of algorithms developed by Marc H. Brown and others at Digital Equipment's System Research Center. It allows users to run an algorithm and to observe it through several views. The construction of views is made easier by the use of a set of graphics and animation libraries. Algorithms and views are separated in the system and so can be built and tested separately. Interesting events are described through an Event Description Language. All client code as well as the system itself is implemented in Modula-3, and objects, strong-typing, and parallelism are used extensively in the system.

2 Overview of Agat

Although algorithm animation systems are becoming more powerful and easier for programmers to use, the task of using these tools to create effective dynamic visualizations of algorithms still involves excessive work. Libraries like XTango provide a powerful two-dimensional graphics package to build very sophisticated animations but require significant and sometimes rather complex additional programming to describe the animation. In systems like Zeus, the client code to be animated has to be implemented in a particular programming language; thus making the animation of an existing program a bit tedious to write.

The Agat system described in this paper provides support for watching a program in action, through the use of a stream processor. The programmer animating a program provides a description of the program's important values using *streams*. Streams can be combined to make other streams. The output of a stream can be piped onto the input of a graphical operator to produce a display. Whenever an *interesting* value is sent on a stream by the program, the value is propagated along all the combined streams, and each stream output updates its visual display appropriately.

Another interesting feature of Agat is the client-server architecture it is based upon: the animation server is a separate process. Only a few lines of code have to be added to the source program in order to animate it. The animation itself is described within an animation file to be loaded in the stream processor (see Fig. 1). This architecture makes the description of the animation independent from the program to be animated. This eliminates the need to rewrite an existing program in a particular language just to animate it and allows the animation to be changed without modifying the source program.

3 Using Agat

We will present the use, functionalities and architecture of Agat on a very simple example, a program written in C to investigate the behavior of a random number generator.

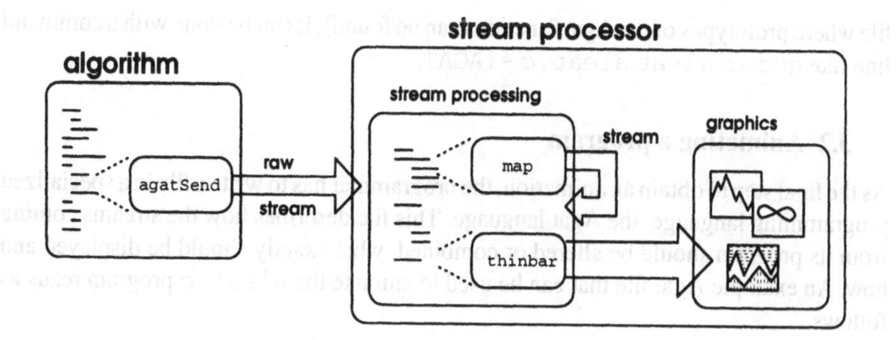

Fig. 1. The Agat architecture

3.1 Preparing a program for animation

To prepare a program for animation, the programmer uses the functions in the Agat library to send the "interesting" data manipulated by the program on streams. The programmer is free to use as many streams as he wants and to name them as suits him (a stream is just identified by a character string).

In our simple example, the program loops, repeatedly sending a new random number on a stream that we choose to name `i`. This is done by calling the `agatSendDouble` function (as the random numbers are of type `double`). The instrumented code reads as follows.

```
/*aleat. c, a pseudo random number generator, instrumented*/
L/* for agat */
#include <stdio.h>
#include <math.h>
#include "libAGAT.h"
double          drand48(void);
/* modify this function to change generators */
double          myRand(void)
{
  return drand48();
}
int             main(int argc, char **argv)
{
  int             i, j, nbn;
  double          v;
  nbn = atoi(argv[1]);
  for (i = 1; i <= nbn; i++){
    v = myRand();
    agatSendDouble("i", v);
  }
}
```

The user needs then to link its program with the Agat library, where the function `agatSendDouble` is defined (note that we had also included the `libAGAT.h` header file where prototypes of the Agat functions can be found). It can be done with a command line like `gcc -o aleat aleat.c -lAGAT`.

3.2 Animating a program

As the final step to obtain an animation, the programmer has to write a file in a specialized programming language, the Agat language. This file describes how the streams coming from its program should be altered or combined, what exactly should be displayed and how. An example Agat file that can be used to animate the `aleat.c` program reads as follows.

```
-- file animAleat.agt
include "stat.agt","misc.agt";-- include some predefined agat
                                     functions
let ci=account(i);  -- count the number of values emitted
                       on stream "i"
let avi=acaverage(i); -- average value on "i"
let inci=increase(i); -- increment between two values
-- we "delay" the streams to be able to have a value
   and the preceding one together
let di=delay(i);
let dci=delay(ci);
let davi=delay(avi);
let dinci=delay(inci);
-- and then we visualize all the streams...
-- a histogram view of "i"
thinbar(i);
-- various plots (points in the plane)
plot(ci, i);
plot(ci, avi);
plot(di, i);
plot(davi, avi);
plot(dinci, inci);
```

Here, from the raw stream `i` of the random numbers, we choose to compute (construct) three other streams. `ci` is a stream counting the number of random values produced (it will thus contain 1, 2, 3, 4, ...). The stream `avi` will contain the successive averages of the generated numbers and `inci` the differences between two successive random numbers.

As we are interested in comparing the current value on these streams with the preceding one, we "delay" them, creating three other streams, `di`, `dinci`, and `davi`. The values on the stream `di` are the values on `i` with a delay of one value. There is a natural notion of time in Agat. A tick of the clock corresponds with the arrival of a value emitted by the program. The streams are therefore naturally synchronized through this clock. This means that, if `i` is 0.2 at time t_0, 0.3 at t_1, 0.1 at t_2 and so on, `di` will be 0.2 at time t_1, 0.3 at t_2... (and will contain no value at t_0).

The functions `account`, `delay`, `increase`, and `acaverage` are written in Agat and defined in the library files `stat.agt` and `misc.agt` included at the top of the Agat program with the `include` directive.

For the animation itself, we choose to plot a histogram of the random values produced (the rendering of the `thinbar` primitive looks like the one of `xload` program) and several simple plots. `plot(ci,i)`, for instance, produces a point for each pair of values arriving on the streams `ci` and `i` (a point for each pair of `(ci, i)` values). All these graphical operators dynamically choose the right scale and ranges and can display axes with smart tick marks. There also exist operators that link successive points with lines, display boxes of varying height and width, and show the ratio of the values on different streams.

The animation is then obtained by running the `aleat` program under the control of the Agat animation server. The simple command line: `agat -f animSimple.agt aleat` creates six windows associated with the six graphical operators called in the Agat program. Figure 2 shows such an animation. Figure 3 shows what is obtained if we replace the classical generator used in our `aleat.c` program (`drand48`) with another one (based on the linear congruential method). The animation demonstrates that this algorithm produces a less random sequence of numbers: there are some discernable patterns.

This simple example illustrates the two most interesting features of our animation tool. First, animating an existing program is very simple: one just includes calls to the

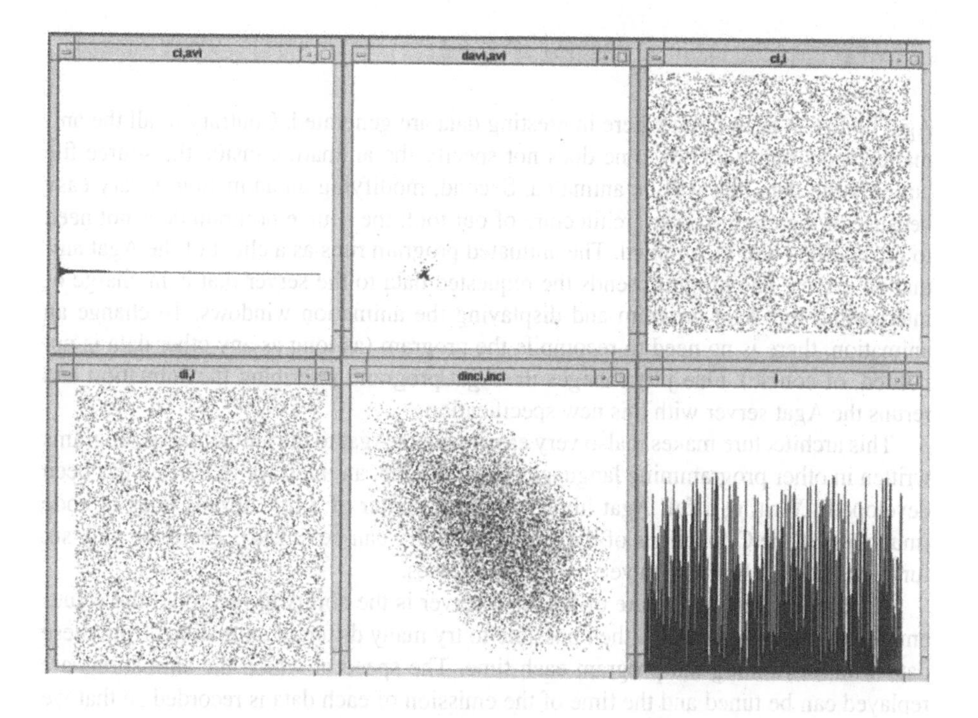

Fig. 2. The result of an animation

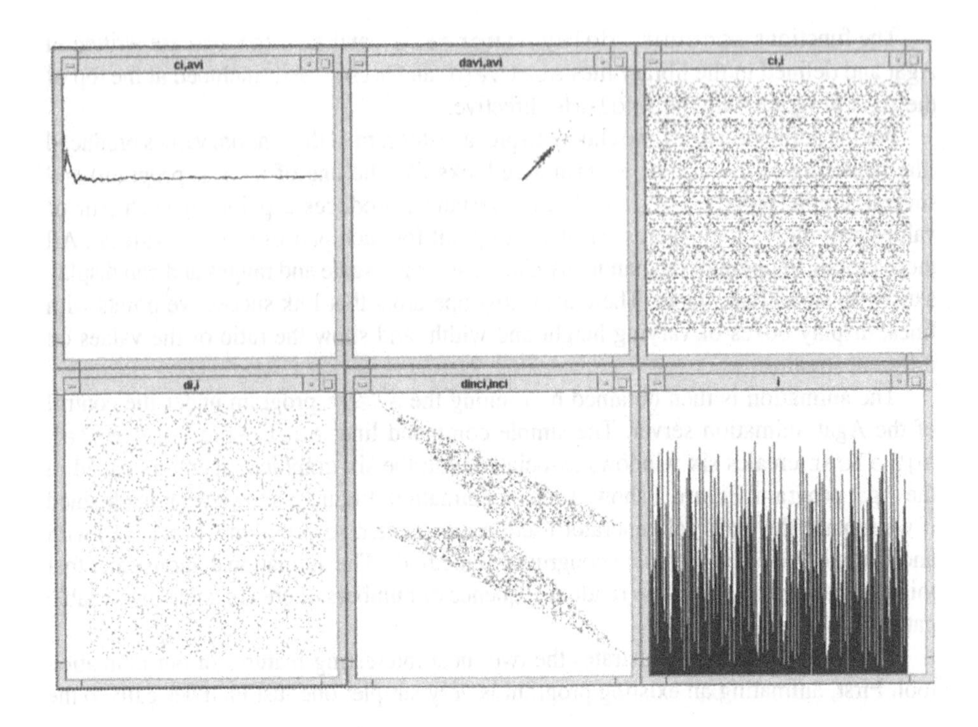

Fig. 3. A "bad" generator

Agat library at each point where interesting data are generated. Contrary to all the animation tools that we know, one does not specify the animation inside the source file but only the data that will be animated. Second, modifying an animation is very easy because of the client-server architecture of our tool: the source program does not need to be modified and recompiled. The animated program runs as a client of the Agat animation server. The program sends the requested data to the server that is in charge of interpreting the Agat program and displaying the animation windows. To change an animation, there is no need to recompile the program (as long as any other data is not needed, of course). One just changes the Agat program describing the animation and reruns the Agat server with this new specification.

This architecture makes it also very easy to adapt Agat to the animation of programs written in other programming languages (for example, an interface to Fortran has been developed). Writing a full Agat library is just a matter of a few dozens lines of code (much less if the C functions of the provided library can be linked). There are only six functions in the library, with very simple interfaces.

Another interesting feature of the Agat server is the ability to record all the values emitted by the program. It is then possible to try many different animations from these data without running the program each time. The speed at which the animations are replayed can be tuned and the time of the emission of each data is recorded so that we can keep a totally accurate image of the progress of the algorithm.

3.3 How to instrument the source code

There are some simple guidelines on how to instrument the code that should be followed to make the best use of the features of Agat. Contrary to most algorithm animation systems Agat enables the user to make many experiments and try various animations without recompiling its program at all. To be as flexible as possible, there are three basic rules:

- only raw data should be sent (do not process data just for sending);
- use a different stream for each different logical information;
- do not hesitate to send more data (use more streams) than intended to show.

As simple processings can be done at the Agat level (the most common mathematical functions are available in Agat) there is no need to do these processings in the program itself where a compilation would be needed to change them. The use of several streams is strongly advised. Two streams can easily be combined into one in Agat. Sending more data than actually needed is harmless but adding a new stream will cost a new compilation. Ignoring a whole stream is very easy: you just don't mention it in the Agat program.

Note that including calls to Agat library functions in the program does not modify its behavior. It can be run without the control of the Agat server and the results will be similar to what would be obtained without the calls to the Agat library. There is, of course, a small runtime overhead. The size of the program is barely changed because the code of the animation is in fact in the Agat server.

4 The Agat language

As seen in the preceding section, to animate a program with Agat, the user has to go through three different steps:

- add "send" commands into the program, to create raw streams of interesting values;
- describe an animation using a configuration file, to build new streams connected with graphical display operators;
- invoke the animation.

The first step is important, because the choice of input data strongly influences the message that the animation conveys. The second step is the most difficult part because the user has to decide what the view should look like in order to highlight interesting information. A view results of the connection of a stream with a graphic display. An interesting point with Agat is that the user can elaborate multiple views from a few basic streams. These views can be changed with no modification of the animated program. The Agat system offers a great variety of constructs to combine raw streams of values into new streams. These constructs are described with a small but powerful language which allows to express subtle mappings on streams, the Agat language. This programming language is quite unconventional. Its main characteristics are the following.

It basically manipulates streams that are sequences of values [see Abelson et al. (1985), for an introduction to programming with streams, and Waters (1991), for a more complete treatment].

Several streams can be combined into a new one with the `map` construct. Various activation patterns can be used to control the merging. The streams are all synchronized with respect to the values coming from the animated program.

The only global variables denote streams and functions on streams. Identifiers that denote elementary values (integers and floating-point data) are local to a `map` construct.

A simple Agat declaration is the one that defines a new stream using a predefined function, and gives it a name with the `let` construct, like in `let ci = account(i)`. The values on the stream `ci` are the successive numbers of values emitted on the stream `i` (that is 1, 2, 3, 4, ...). The predefined function `account` is actually written in Agat:

```
let account(i) =
      map{c:=0}(_ -> c:=c+1)i;
```

The definition of `account` uses the basic construct `map` (with a register declaration and a wild card pattern that will be explained later). Basically, `map` constructs a new stream by applying a transformation to all the values of its input stream. For example, given a stream `f`, then `let logf = map(x -> log(x)) f;` defines a new stream `logf` of the logarithms of the values of `f`. This means that each time a value `v` is emitted on the stream `f`, the value `log(v)` is emitted *at the same time* on the stream `logf`. We will give later more details on what "at the same time" means. The operator `map(x -> log(x))` maps the mathematical function `log` on each value of the stream it is applied to. The left side of the arrow is called the activation pattern and the right side is the action to be performed on the values.

4.1 Activation patterns

The activation pattern describes the state of the input stream(s). For a given stream, this state could be one of the following:

- *new value*: a new value is emitted on/by the stream (actually, the propagation of a value on a stream is instantaneous in Agat);
- *no value*: no (new) value has been emitted on the stream (older values may have been already emitted);
- *old value*: no (new) value has been emitted on the stream but at least one older value was;
- *not yet a value*: no value at all has ever been emitted on the stream;
- *first value*: a new value is emitted on the stream and it is the first one.

Activation patterns are very useful for mappings with multiple inputs. For example, the stream `h` defined by `let h = map(a,b -> a+b) f,g;` maps the function + on the two input streams `f` and `g`. The simple activation pattern `a,b` implies that the values emitted on `f` and `g` are *buffered* until there are two new values that are available on both `f` and `g`.

Let `hh` be the stream defined by:

```
let hh = map(a,b -> a+b
                |a,* -> a
                |*,b -> b)f,g;
```

Table 1

Stream	Time →					
f	1		2	3		4
g		10	20		30	40
h		11	22		33	44
hh	1	10	22	3	30	44
av		5.5	11.0	11.5	16.5	22.0

The star matches the state "no value". If a new value is only emitted on the stream `f` and nothing on `g`, the pattern `a,b` does not match (like in the case of the previous `h` stream), but the pattern `a,*` does. Notice that if `f` and `g` are two raw streams (streams that come directly from the program), the pattern `a,b` can never succeed as two values are always emitted at a different time, and `hh` is just a merged stream of `f` and `g`, containing the values of `f` and `g` in the order where they have been emitted.

Another example of a map with multiple inputs and special activation patterns is given by the stream `average` defined by:

```
let average = map(a,b  -> (a+b)/2
                 |a,b+ -> (a+b)/2
                 |a+,b  -> (a+b)/2)f,g;
```

Given the streams `f` and `g`, the stream `average` carries the average values of the two more recent values emitted by `f` and `g`. The pattern matching and the corresponding action performed on the inputs of a map construct are *instantaneous*. Thus, the outputs of `average`, `f`, and `g` are *synchronized*, that is, they occur "at the same time". We summarized the behavior of the streams `h`, `hh`, and `average` (`av`) in Table 1. The values in columns are synchronized.

4.2 Registers

Introducing a new map construct by registers, the following defines the stream function `delay` that creates a delayed copy of a stream:

```
let delay(i) =
        map{last,tmp}
            (x^-> last:=x, null
            | x-> tmp:=last,last:=x,tmp)i;
```

Notice that `delay` is not a stream, but a *function* that takes a stream as input and returns a new stream. The hat matches the state "first value". The identifiers appearing inside brackets are called registers. They act as local variables for the map construct they appear in. When a sequence like `tmp:=last, last:=x, tmp` occurs, each instruction is evaluated and the last expression is the value output by the stream. The `null` instruction is used when there is no value to output. An important rule is that inside a map, the patterns are checked in the order they appear. Hence, in the example of delay, the first value emitted by the stream parameter i matches the first pattern and initializes the local variable last to its proper value. In addition to the predefined functions (like +) and the assignment (:=), the Agat language provides a conditional statement. The stream function positive creates the stream of the positive values of a given stream:

```
let positive(i) =
        map (a -> if a>0 then a else null) i;
```

As registers allow to build *accumulators*, conditional statements allow to create *filters*. Mapping, filtering, and accumulating are the basic functions required for a powerful stream language (Waters 1991).

Although the user can easily write its own functions, the Agat system comes with a library of predefined stream functions (that can be used via the include statement). For example, the following function extremafilter which extracts the local extrema of a given stream is available in the library:

```
let extremafilter(i) =
    map {slope,last,tmp}
        (x^ -> last:=x, slope:=0, null
        |x  -> if sgn(x-last) != slope
                  then tmp:=last, slope:=sgn(x-last),
                       last:=x, tmp tmp
                  else last:=x, null
                  end ) i;
```

5 The graphical features of Agat

Agat does not possess yet a library of very sophisticated graphical operators. In fact the library is rather poor, compared to more sophisticated algorithm animation systems such as Zeus. For now, the strongest points of Agat are its ability to mix and transform streams and its flexibility (changing animations without recompiling and even without running the program again).

The graphical operators that have been implemented can be classified as follows:

- histogram operators can plot one or more streams as histograms (different streams are assigned different colors);
- ratio operators only show the ratio between the values from two or more streams (by filling a rectangle with different colors);
- plotting operators basically plot points in the plane (the coordinates are taken from two streams), consecutive points can be linked with segments;
- box operators plot rectangles in the plane (that can be filled with a chosen color) whose heights and widths are defined on the streams;
- matrix operators are designed to display matrices (they are displayed as grids and a color can be assigned to a range of values);
- three-dimensional plotting operators that plot a function of two values.

All graphical operators can display the coordinate axes (with or without tick marks) and the scale is dynamically changed to fit in the animation window. Some operators can erase previous drawings. This can provide a simple way to move objects around. It is also possible to generate PostScript code to save the drawing in an animation window.

Plotting operators provide two interesting interactive features: the user can zoom on a part of an animation window (obtaining a new animation window for the selected zone) and ask for the coordinate of a point with the mouse. 3-D graphics can be rotated interactively with the mouse to change the view point.

6 Some applications of Agat

Many algorithms of computer algebra are known to be quite complex and to exhibit a complicated behavior at run-time, very dependent on the input data (and not only on their sizes). Algorithm animation could be a very useful tool to monitor various quantities in these algorithms and thus to help in their understanding, debugging, and optimization.

Agat has been used in the SAFIR project (the symbolic computation group at INRIA) for animating several complicated algorithms related to computer algebra. Agat has been used to investigate Gröbner basis algorithms, showing the growth of the number of critical pairs with different strategies and monitoring the memory consumption.

Another application was the design of microwave filters where transfer functions of very high degree occur, causing various numerical problems. Agat was used to evaluate different methods.

Figure 4 a–d shows four images taken from an animation that lasts a few hours of real time. These images describe the behavior of a variable-step-size integrator using a Newton method. For example, the peak in Fig. 4 c indicates that the algorithm has done something wrong. In Fig. 4 d, the flatness before the peak indicates that the integrator has basically run for nothing, reaching the machine accuracy. This kind of property is very difficult to catch without a dynamic graphical representation.

In this case, the ability to record and replay the animation at different speeds was crucial to gain some understanding of what is going on. In fact, this example generates several megabytes of data and leads us to implement the compression of the saved values on the fly.

Agat was also used to instrument a program that studied plane complex curves and their singularities (de Sousa 1995). The implemented method is a mixed numeric and symbolic method. It was an invaluable help for the development to be able to watch the progress of the algorithm in real-time, detecting numerical problems (including numerical instabilities) as well as symbolic inconsistencies (a wrong degree for an intermediate polynomial or a wrong number of roots). Figure 5 a and b is the result of an animation of this program. It shows objects related to the computation of toric knots (knots on torus) that describe topological invariants of a singularity.

7 Implementation

Agat is implemented in C. The current version of Agat uses the ASAP protocol – a protocol to exchange mathematical objects (Dalmas et al. 1994) for transmitting data between the client program and the animation server. The Agat library is around 500 lines of code. The animation server is 8000 lines of code for the Agat language interpretation part and 6000 lines for managing the graphics. The animation server works by transforming the Agat program into a graph that expresses the dependencies between the streams that are

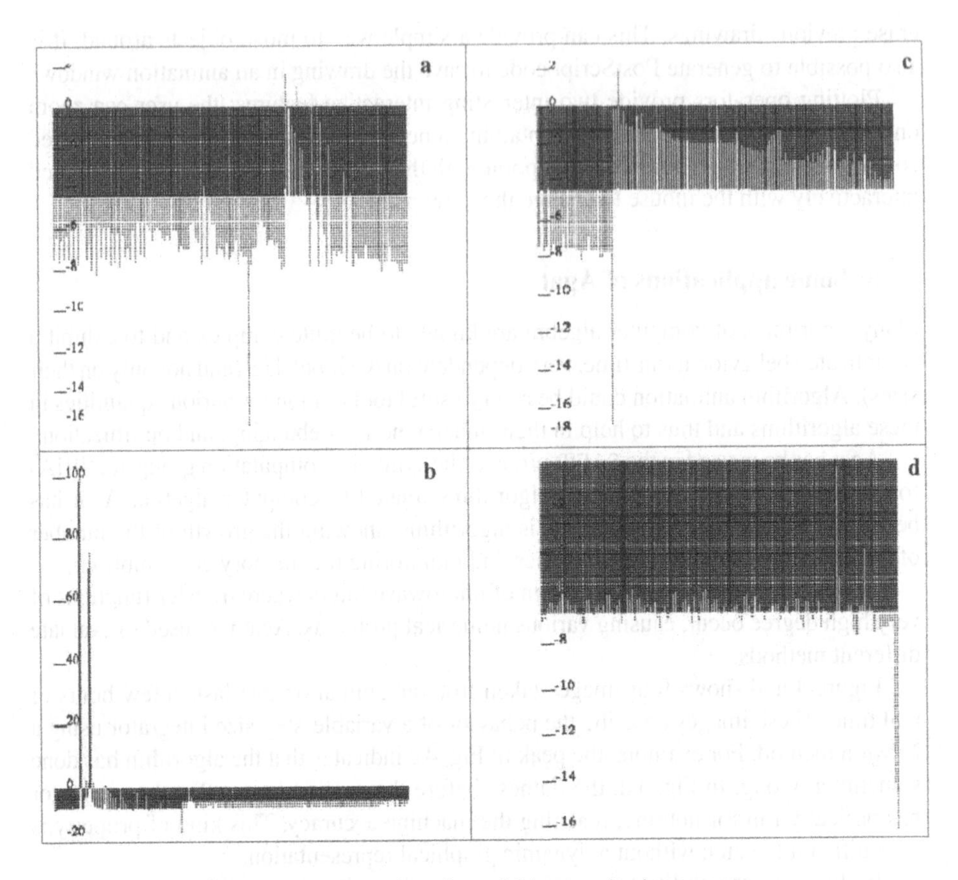

Fig. 4 a–d. Behavior of a variable step-size integrator

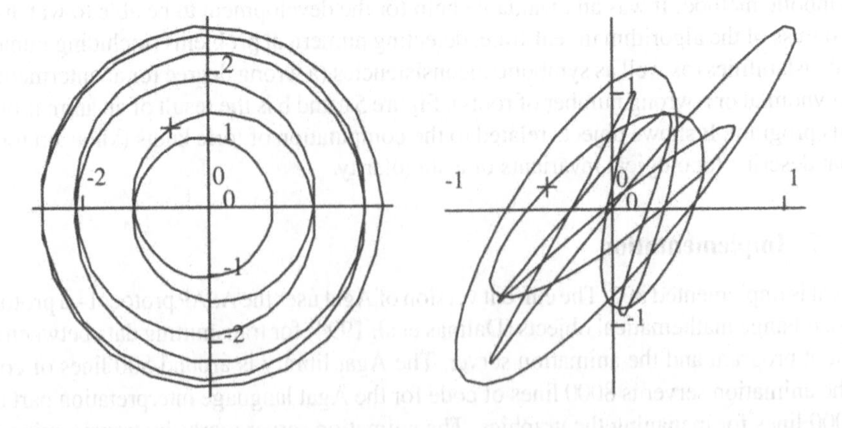

Fig. 5. Topological invariants of a singularity

used. This graph is then interpreted: each time a new value arrives from the animated program, it is propagated along this graph as far as possible, triggering activation patterns and associated computations or being stored in registers, eventually ending in a graphical operator.

Adding a new graphical operator to the Agat animation server is an easy task (the hard part is writing the operator itself). The interface between a graphical operator and the server is very simple.

Agat is available by anonymous FTP at zenon.inria.fr in the safir directory. The current version has been extensively tested on SparcStations running SunOS4.1.x and DECStations running ULTRIX and should run on most workstations that have X11 available.

8 Conclusion

Agat has been successfully used in our research group to investigate the behavior of complex algorithms and also as an occasional debugging tool as well as an interactive (on-line) plotter. For example, a binomial Gröbner basis algorithm has been studied as well as a very complex identification problem involving high-degree polynomials for the design of microwave filters. For our uses, Agat could not have been replaced by any other more sophisticated animation tool. They would have required too much efforts to set up interesting animations for our colleagues to accept them. The graphical possibilities of Agat can be considered rather poor for an algorithm animation system but we found that looking at several simple animations in parallel can often give more insights than looking at a single sophisticated view.

From a language design point of view, some people will object that we certainly don't need another programming language, even for algorithm animation. When we designed the language, we had considered extending an existing "little" language (like Scheme or TCL). Two main reasons made us choose to design yet another language. The first is that we did not see how the main features of Agat, streams processing with pattern-matching and synchronization, could have been nicely integrated in such a language (some difficult semantics issues arise that we can control in the context of the limited constructs of Agat). The second one is the desire to keep the language as simple as possible to be accessible to the largest possible class of users. For the moment, learning Agat has never been a problem for our users (even for rather "math-oriented" people). Of course, in the long term, the risk is that the needs to extend Agat could reach the point where it should become a real full-featured programming language. This could force us to reconsider our decision but we hope to be able to carefully control the evolution of Agat to keep it as simple as possible.

In the near future, we plan to add more sophisticated graphical operators, including 3-D graphics. For now, Agat has only basic three-dimensional plotting operators. 3-D graphics can provide richer visualizations of arrays. There is certainly a lot to do with 3-D graphics for algorithm animation, in the spirit of Brown and Najork (1993) and Stasko and Wehrli (1993). Some powerful systems have been developed for interactive, animated 3-D graphics that can be connected to an embedded language and that could thus provide a good basis for rich algorithm animations (Najork 1994). Considering our approach of providing relatively simple graphical operators, it is not clear that going from 2-D to 3-D would really be a very valuable improvement in the sense that the use of these new functionalities could be rather low, the simple things being often the most useful. We are also investigating the possibility of adding primitives to create and move around simple objects, following the work of Krishnamoorthy and Swaminathan (1989). The challenge here is to keep the Agat language, and most interesting Agat programs that should use these new functionalities, simple enough. An almost trivial extension would be the addition of sound operators that would produce sounds instead of graphical actions.

An interesting long-term project can be the embedding of Agat in a source-level debugger like `gdb`. The basic idea is to offer full access to all the Agat functionalities to animate the evolutions of the interesting data of a program running under the control of the debugger. The Prism programming environment of Thinking Machines Corporation for Connection Machine systems already includes a debugger that has the ability to visualize arrays in several graphical ways but this cannot be considered as a real animation with the possibilities of Agat. Another (easier) idea could be to turn the Agat server into a real interactive interpreter where the user could stop the computation, introduce new animation code, and restart the program at will. This can be considered as a nice extension of the "debugging by printing" method.

Even if Agat can only deal with numerical values, but not complex symbolic objects, we think that it could be a very effective tool to look at complex symbolic algorithms (as exemplified by the first experiences in our research group), because, after all, the evolution of simple numerical data is all that most people can easily handle (Bertin 1983).

Acknowledgments

We would like to thank Stéphane Lavirotte who implemented the 3-D operators and our first users in the safir project who help us to debug and enhance Agat.

References

Abelson, H., Sussman, G. J., Sussman, J. (1985): Structures and interpretation of computer programs. MIT Press, Cambridge, MA.

Bentley, J. L., Kernighan, B. W. (1991): A system for algorithm animation. Comput. Syst. 4/1: 5–30.

Bertin, J. (1983): Semiology of graphics. University of Wisconsin Press, Madison, WI.

Brown, M. H. (1988): Exploring algorithms using BALSA-II. IEEE Computer 21/5: 14–36.

Brown, M. H. (1992): Zeus: a system for algorithm animation and multi-view editing. Digital SRC Res. Rep. 75, DEC, Palo Alto, CA.

Brown, M. H., Hershberger, J. (1991): Color and sound in algorithm animation. Digital SRC Res. Rep. 76a, DEC, Palo Alto, CA.

Brown, M. H., Najork, M. A. (1993): Algorithm animation using 3D interactive graphics. Digital SRC Res. Rep. 110a, DEC, Palo Alto, CA.

Brown, M. H., Sedgewick, R. (1984): A system for algorithm animation. Comput. Graph. 18: 177–186.

Cox, K. C. (1992): Abstraction in algorithm animation. In: Proceedings of the 1992 IEEE Workshop on Visual Languages. IEEE Computer Science Press, Los Alamitos, CA, pp. 18–24.

Dalmas, S., Gaëtano, M., and Sausse, A. (1994): ASAP: a protocol for symbolic computation systems. Rapp. Tech. 162, Institut National de Recherche en Informatique et en Automatique, Le Chesnay.

de Sousa, J. (1995): Etude algorithmique de la topologie des courbes algébriques planes complexes. Ph.D. thesis, Université de Nice-Sophia Antipolis, Sophia Antipolis, France.

Duisberg, R. A. (1986): Animated graphical interfaces using temporal constraints. In: Proceedings of the ACM CHI '86 Conference on Human Factors in Computing Systems. Association for Computing Machinery, New York, pp. 131–136.

Gloor, P. A. (1992): AACE algorithm animation for computer science education. In: Proceedings of the 1992 IEEE Workshop on Visual Languages. IEEE Computer Science Press, Los Alamitos, CA, pp. 25–31.

Helttula, E., Hyrskykari, A., Räihä, K.-J. (1989): Graphical specification of algorithm animations with Aladdin. In: Proceedings of the 22nd Hawaii International Conference on System Sciences, pp. 892–901.

Krishnamoorthy, M. S., Swaminathan, R. (1989): Programs tools for algorithm animation. Software Pract. Exper. 19: 505–513.

Najork, M. A. (1994): Obliq-3D tutorial and reference manual. Digital SRC Res. Rep. 129, DEC, Palo Alto, CA.

Rasala, R., Prouxl, V. K., Fell, H. J. (1994): From animation to analysis in introductory computer science. SIGCSE Bull. 26/1: 61–65.

Stasko, J. T. (1990): Tango: a framework and system for algorithm animation. IEEE Computer 23/9: 27–39.

Stasko, J. T. (1992): Animating algorithms with XTango. SIGACT News 23/2: 67–71.

Stasko, J. T., Wehrli, J. F. (1993): Three-dimensional computation visualization. In: Proceedings of the 1993 IEEE Symposium on Visual Languages. IEEE Computer Science Press, Los Alamitos, CA, pp. 100–107.

Tal, A., Dobkin, D. (1994): GASP: a system for visualizing geometric algorithms. In: IEEE Visualization '94, pp. 149–155.

Thinking Machines (1994): Prism user's guide, version 2.0. Thinking Machines Corp., Cambridge, MA.

Waters, R. C. (1991): Automatic transformation of series expressions into loops. ACM Trans. Programm. Lang. Syst. 13: 53–98.

Computation and images in combinatorics

Maylis Delest, Jean-Marc Fédou, Guy Melançon, and Nadine Rouillon[1]

1 Introduction

Combinatorics has always been concerned with images and drawings because they give interpretations of enumeration formulae leading to simple proofs of these formulae, and sometimes they are themselves central to the problem. Even if some small example drawings do not contain all elements of the proof, they are often useful to guide the intuition.

In recent years, enumerative combinatorics has developed from this point of view, in interactions with theoretical computer science and graphics computer science (see, e.g., Arques et al. 1989). In the field of bijective combinatorics linked with computer science, the DSV methodology (X. Viennot 1992, Delest 1995) has led to new formulae. The problem often consists in establishing enumerating formulas by constructing bijections between sets of objects having the same cardinality. Computations are then suppressed and are replaced by effective constructions of bijective correspondences between sets. The formulae may then be viewed as reflecting combinatorial properties of the objects involved. The problem may also consist in finding geometrical transformations of objects reflecting the equations which relate them.

For several years already, enumerative combinatorics research uses symbolic computation. Most of the computer algebra systems, including Macsyma (Symbolics 1984), Maple (Char et al. 1992), and Mathematica (Wolfram 1988), allow easy algebraic and graphic manipulations but none of them gives a library of functions working directly and interactively on the graphic representation of the object. Some projects have led to tools for performing such actions. For example, we refer to the Mathematica library done by Skiena (1990) which gives tools for implementing combinatorial objects. This library does not allow to interact with the graphics itself. The aim of the CalICo (calcul et image en combinatoire) project is the development of a software in which the manipulation of combinatorial objects can be made through all their representations: coding, formal power series, and overall pictures.

1.1 A famous example

As an example we discuss the Robinson–Schensted correspondence (see Robinson 1948, Schensted 1961, G. Viennot 1977, for details).

---

[1] When this publication was initiated the contributors to the CalICo project were Y. Chiricota, M. Delest, J.M. Fédou, V. Gaudin, G. Melançon, and N. Rouillon.

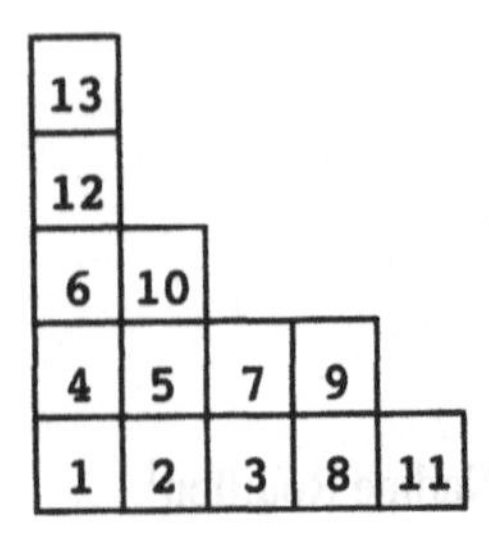

Fig. 1. A standard Young tableau

In the theory of representations of finite groups, there is a formula linking the number of elements of a group G and the dimension of its irreducible representations. The case $G = S_n$, the group of permutations of $\{1, \ldots, n\}$, has been studied in great detail over the years. In that case, irreducible representations correspond to Ferrers diagrams filled with distinct integers to form a *standard Young tableau* (see Fig. 1). (Ferrers diagrams are built by piling rows of cells. Rows are left justified and decrease in size. In a standard Young tableau, labels in rows and columns increase from left to right and from bottom to top respectively.) The equality then reads:

$$\sum_{\lambda} f_{\lambda}^2 = n!,$$

where f_λ is equal to the number of standard Young tableau with underlying Ferrers diagram corresponding to λ, for more information about Robinson–Schensted correspondence in connection with the theory of group representations, see, e.g., Kerber (1991).

Enumerative combinatorics then plays with algebra to give some insight on this formula. The Robinson–Schensted correspondence shows that there is a bijection between permutations and ordered pairs of standard Young tableau. This correspondence is in fact an algorithm taking as input a permutation and constructing a sequence of ordered pairs of tableau by reading the permutation from left to right. Many combinatorial properties of the permutation can be read on the tableau associated with it. For instance, taking the inverse of the permutation corresponds to exchanging tableau. In other terms, if (P, Q) is the pair of Young tableau associated to a permutation σ then (Q, P) is associated to σ^{-1}. This phenomenon has been explained in an elegant manner by G. Viennot (1977). His methodology conforms to the point of view mentioned above and motivates the development of a project such as CalICo.

1.2 A piece of history

CalICo was initiated by M. Delest in 1989. One of the key problems in combinatorial work consists in finding an appropriate object representation, i.e., one which is useful with respect to certain transformations or manipulations of the objects in question. The key idea of CalICo was to provide a computing environment to assist this process in an essential way. For example, one should work with a grid of points instead of the standard representation of a permutation as a word. CalICo was designed to be an experimental laboratory and wishes to offer a graphically-rich environment that facilitates working in enumerative combinatorics. Eventually, it should be a guiding tool for finding bijections between a family of objects. A component of CalICo would gather informations about objects and organize them in order to assist the user. That component would be a tutor (see Sects. 4 and 5) for studies in combinatorics. In 1990 the project entered a concrete stage. After a primitive realization of the tutor and a first realization of a workshop (see Sect. 2) for manipulation and edition of polyominoes, some combinatorists (at LaBRI, University of Bordeaux I) saw immediately how CalICo could be a tool of great help. Some of them joined the project and proposed visualization workshops for other families of objects. A visualization workshop for braids in 1991, and one for permutations in 1992 were added to the preceding workshop for polyominoes. The communication manager was an essential part at the first day. It has also been completed in 1992.

At the beginning of 1993, a second version of the tutor was developed and the visualization workshops were improved. After demonstrations at scientific meetings, the enthusiasm expressed by the scientific community and the growing demand for installing CalICo on many sites gave the project a new impulse. The main concern was then to unify the appearance of visualization workshops, to standardize many of their functionalities and to systematize the development of new visualization workshops. With the help of new members of the CalICo team, the structure of a standard visualization workshop was elaborated; a skeleton including common functionalities was developed. The foundations of CalICo's architecture were set right from the beginning. Because of the many components interacting and communicating with each other, the software may be viewed as having a distributed architecture (see Sect. 7).

1.3 About this paper

The present paper presents the CalICo software with a special emphasis on its user interface and the interaction between images and combinatorial objects. See Rouillon (1994) for a full description of the CalICo package. In Sect. 2, we show an overview of its structure. To guide the reader, we show simple examples. Then we describe the tutor (which concerns combinatorics knowledge) that is the heart of the CalICo environment. Section 5 is devoted to the formal coding and to links with computer algebra systems. In Sect. 6, we describe the primary interface that allows the user to launch an application in the system and the model that we have designed to easily produce new interfaces. Section 7 is devoted to the communication manager.

2 Overview of CalICo

In this section, we describe the general concept of CalICo. CalICo gathers in a single software environment a set of different applications, called workshops. These workshops can be run on remote computers linked by the Internet network. This software environment allows message exchange between each of those workshops. It offers a graphical and mathematical working environment.

We use the term *workshop* instead of the more usual terms *application*, *program*, or *software* because it gives the intuitive notion of workshops working together in a factory. A workshop in CalICo is independent of the others and is dedicated to a specific job. Independence means that each workshop could be used without CalICo. Of course, in this case, the workshop loses some of its functionalities.

Various workshops coexist within CallCo. The tutor is the main workshop. The others could be divided in three groups: visualization workshops, computation workshops, generating workshops. All of them work in a concurrent way. They are gathered together in a distributed architecture. The client/server model is used.

2.1 Workshops

The main workshop: the tutor. The tutor is the heart of the CallCo environment. Its functions are to supervise the system and to guarantee combinatorial coherence. This tutor is actually specific to CallCo (see Sects. 4 and 5). It is exclusively dedicated to enumerative and algebraic combinatorics. The tutor is able to classify mathematical objects according to their combinatorial properties. It offers experimental ways in order to suggest possible one-to-one correspondences between the manipulated structures to CallCo's users.

Visualization workshops. Many researchers in combinatorics handle both mathematical formulae and graphical representations of combinatorial objects. A very enhanced example is in X. Viennot (1988). X. Viennot often uses the phrase:

"Dessiner des calculs, calculer des dessins"

("drawing computations, computing drawings"). CallCo includes a set of workshops which allows visualization and handles various combinatorial structures (see Sect. 6).

Computation workshops. As we saw before, it is not enough to manipulate graphical objects. Users need also to access them by the mean of their formal coding. So, to be complete, we have to associate computational workshops with graphical workshops. This will allow the manipulation of formal coding of objects (see Sect. 5). For the moment, CallCo integrates two packages, Maple (Char et al. 1992) and Mathematica (Wolfram 1988).

Generating workshops. To compute significant statistics (or properties) on objects, people often need to consider large random objects. So, CallCo includes generating workshops. In Sect. 6 we present algorithms used in some of these generating workshops.

2.2 Collaboration between workshops

The ultimate goal is to produce a system that can be operated easily by a non-computer scientist and to use methods leading to easy extensions and generalizations of already existing modules.

On the one hand, we group (see Sect. 7) in a single software environment all the workshops we need. We propose a user-friendly graphical interface to allow communication between workshops.

On the other hand, a *distributed architecture* has been chosen. Each workshop can be run on a remote computer. This distributed aspect allows parallel treatments of several processes. Advantageously, generating workshops run on different computers in order to increase the speed of the system.

Workshops could also be dynamically visualized on a remote display. The benefit is twofold. First, one can use several terminals or computers to display one's work. Second, this opportunity allows us to visualize the result of a computation (by way of a drawing with colors) on the display of a person in another city or another country. This last notion is very close to the groupware field (Beaudoin-Lafon and Karsenty 1992). In CalICo, remote display is not based on the functionalities provided in standard by X11; instead, only the data giving the coding of the objects are transferred between two remote applications (see Sect. 7).

From a technical point of view, the implementation is based on the *client/server model.* GeCI (Gestionnaire de Communications Interactif, interactive communications manager) is a unifying program. It manages dynamic connections between programs (even on remote computers) and the exchange of messages between them (see Sect. 7 for more details). Here, we would like to highlight a particular point: if the data are on the same display they are transmitted by a copy/paste operation even if the workshops are running on different computers.

2.3 Configuration and extensibility

CalICo can be configured at different levels. Some configurations can be done by a simple user, others require more knowledge about computer systems.

Dynamic configuration for simple user. Using the mouse, during a CalICo session the user can run (or kill) as many workshops occurrences chosen among available workshops as he wants. The user can run (and visualize) these workshops on any computer on the Internet network registered in his installation of CalICo (see below). We can say that the user creates a "work space on the net" and builds a *virtual computer* with which he will work.

Personal configuration for simple user. This kind of configuration only requires to learn the configuration file syntax. When CalICo is launched, its *configuration file*, called `.CalICo.config`, is parsed. This file completely describes the *piloting interface* (see Sect. 6), e.g., the computers and displays allowed for this user.

New workshop for programmers. Accessing a new workshop does not need special knowledge. It consists in adding a few lines to the configuration file of CalICo. On the contrary, the conception of a new workshop is a programmer job. Still, his work will be greatly facilitated by the use of the CalCom library (CalICo 1993) and the CalICo model workbench (Gaudin 1995).

New combinatorial knowledge for Maple users. The user adds a combinatorial knowledge to the tutor by means of a graphical interface. The user has to know the *formal coding* of the object he wants to manipulate (see Sect. 5) as well as CalICo's terminology. To add significant information, for example a predicate, it is better to know Maple (which is the programming language of the tutor) and combinatorics.

Below, we discuss more precisely all those various CalICo aspects.

3 The use of CalICo as illustrated by examples

When one issues the `calico` command at the top-level shell, its *configuration file*, which name is `.CalICo.config`, is read by a parser and a scanner (produced by means of the GNU software, Flex and Bison). This file has to be found in the user's home directory in the directory `~/.CalICo`. If it is not found there, a file by default is copied from the system. This file contains the description of the available workshops and therefore describes exhaustively what we call piloting interface for CalICo (see Fig. 2 a).

On the left side of the window of the piloting interface, one finds all available

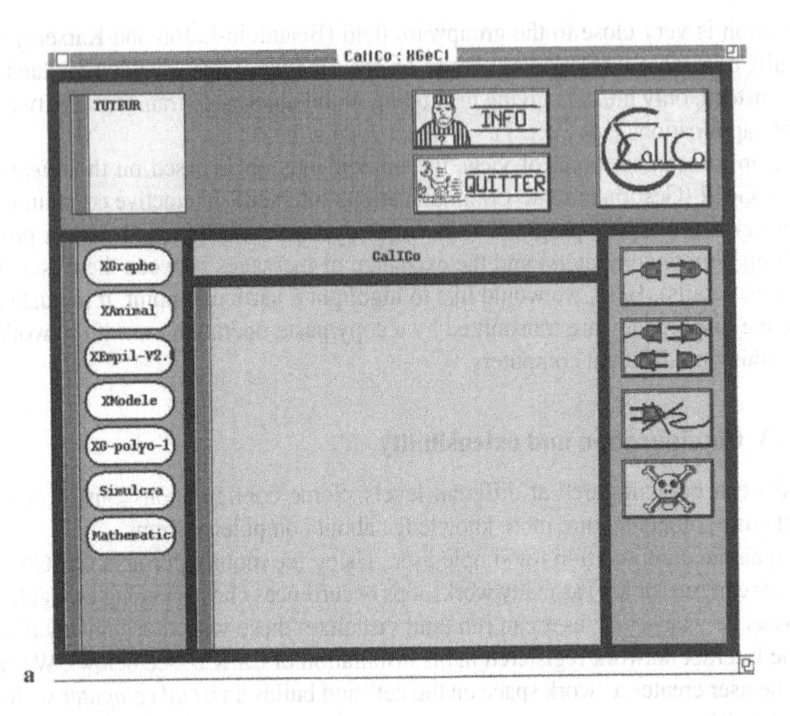

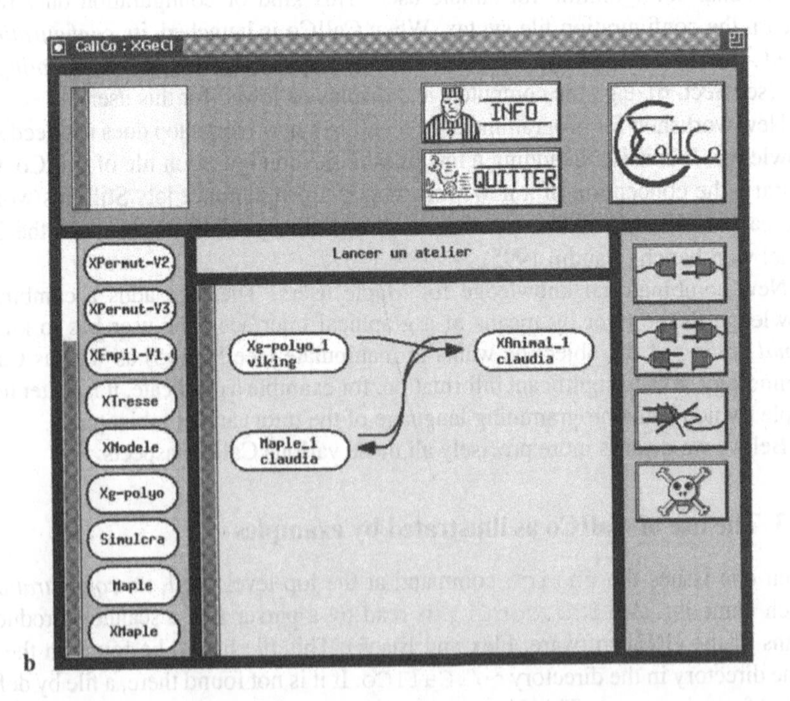

Fig. 2. Piloting interface **a** at the starting up, **b** after connections

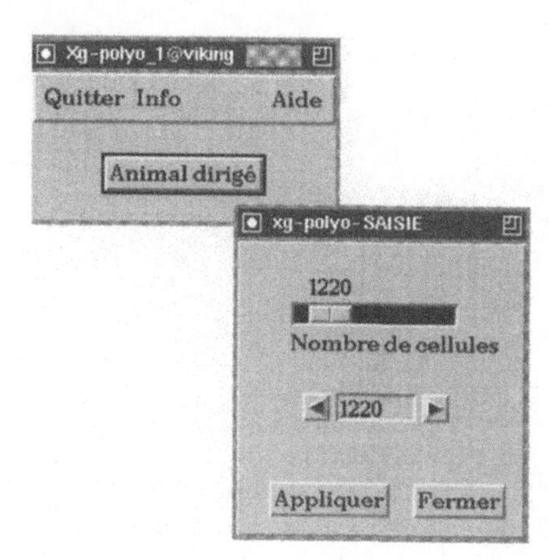

Fig. 3. Random generation of animals

workshops. On the right side, the buttons allow to connect the output of a workshop to the input of another and to destroy a launched workshop. As an example, one can launch successively the three workshops, *Xg-polyo, XAnimal*, and *Maple* and connect *Xg-polyo* to *XAnimal* and *XAnimal* to *Maple* (and reciprocally). The result is shown in Fig. 2 b.

When the user launches a workshop, a window gives her the possibility to change the name of this workshop occurrence as well as the computer used for running and display. If the user chooses a remote machine for the execution, it is necessary that the CalICo daemon process, called *calicod*, runs on this machine. Starting this daemon can be done by any user and requires only to type the `calicod` command at the shell level. On a workstation, the user must remember to add this host to the list of machines that are allowed to connect to the local X server. Use server access control program for X and execute `xhost <hostname>`.

For example *Xg-polyo* (Fig. 3), generator for directed animals, was launched on a remote machine (here "viking", the local machine being "claudia" as shown in Fig. 2 b).

The motivation of the configuration in Fig. 2 b can be found in Denise and Rouillon (1992).

Let us consider another situation when one wants to test some Maple functions on a heap of pieces (X. Viennot 1986). In this case, we need a big number of input/output moves between the manipulation of heaps in the graphical interface and Maple. Here one may prefer to start *Xmaple* and take advantage of the graphical user interface of Maple directly. Exchanges can then be made by *copy/paste* using the mouse (left button) at once with the CONTROL key.

In the workshop *XEmpilement*, by a CONTROL/left-button action in a heap (Fig. 8 a) one can select a copy of the formal coding underlying the heap. In *XMaple*, a middle-button action pastes the coding as observed in Fig. 7. Reciprocally, in *XMaple*, after application of the function f on the heap, using the mouse one can copy the consequent coding. If one pastes (clicking the mailbox icon with the middle button) in the workshop, the formal coding is received and then translated. A new heap is drawn (Fig. 8 b).

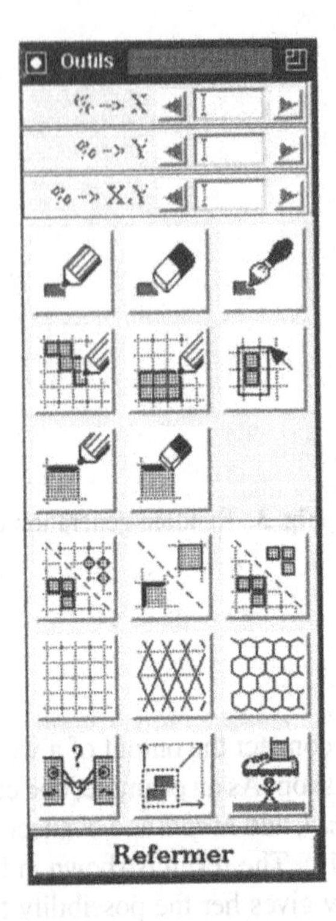

Fig. 4. Toolbox of animals workshops

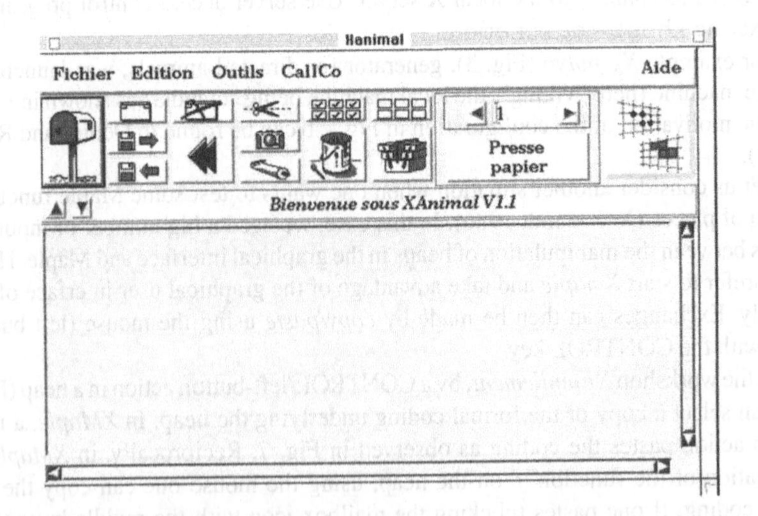

Fig. 5. Animals and polyominoes visualization

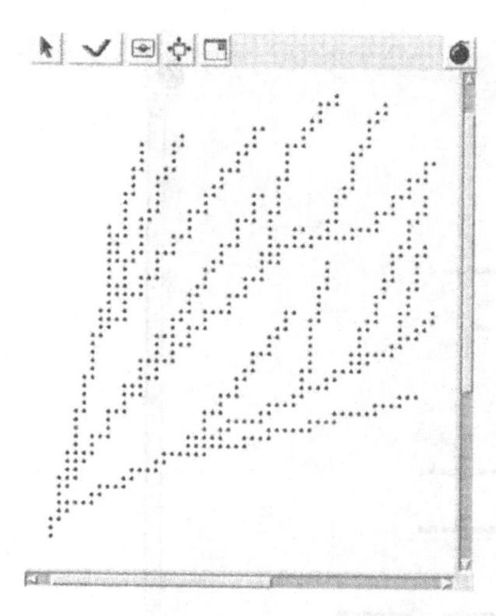

Fig. 6. A random directed animal

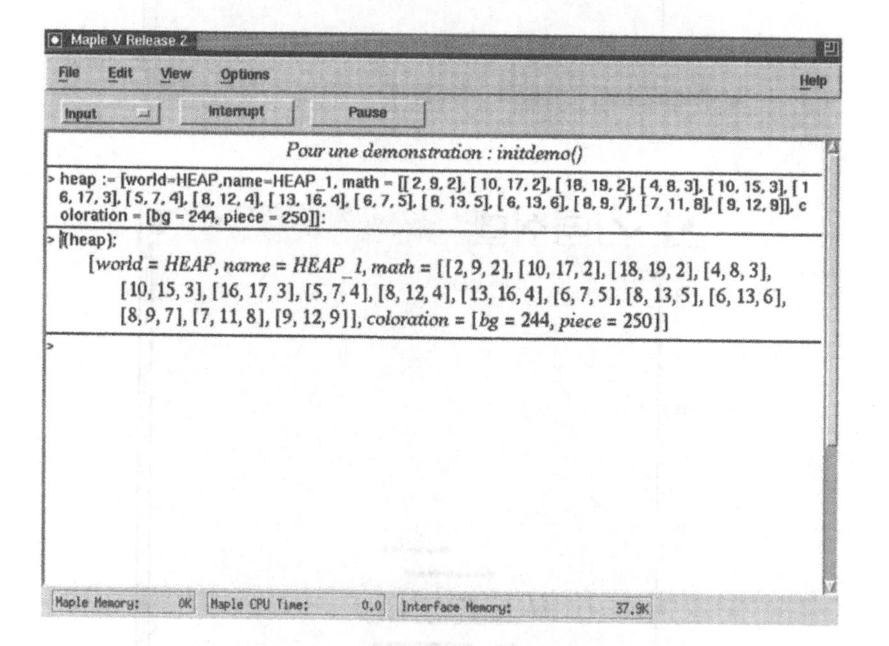

Fig. 7. *Xmaple* session, with paste and copy

4 The tutor

CallCo's tutor is intended for the manipulation of any combinatorial objects. Its two main purposes are to classify objects by an automatic process and to use colors on objects in order to obtain a symbolic representation of their combinatorial properties.

First of all, we define a terminology well adapted to CallCo. We show the inheritance relationships between manipulated structures. We present the dependency graph

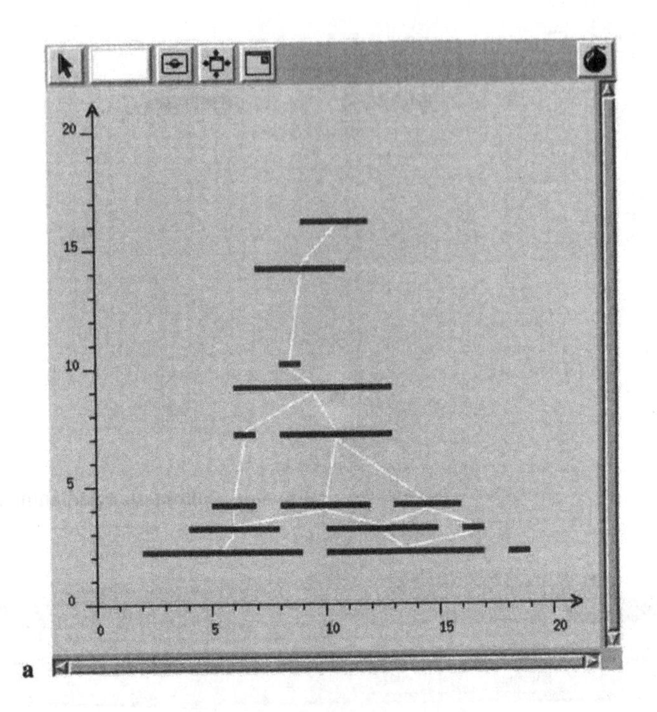

a

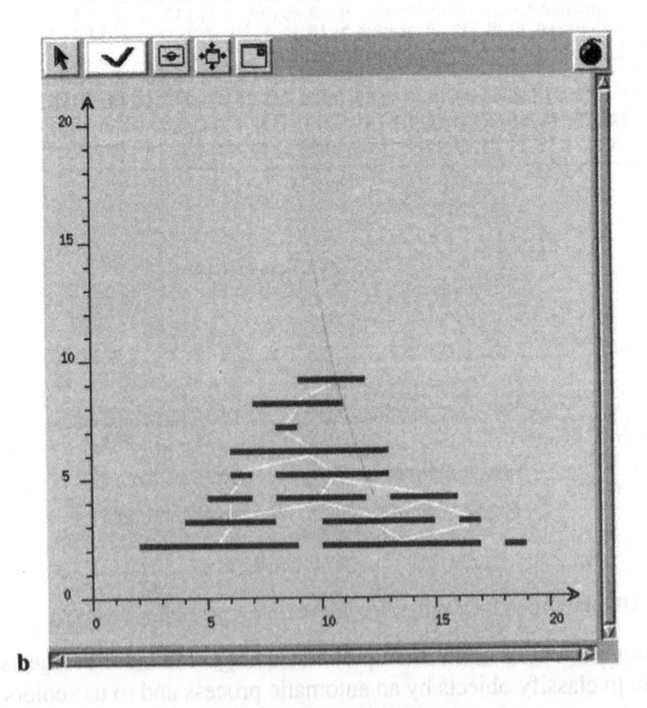

b

Fig. 8. a Initial heap, called h; **b** pasted heap after the computation of $f(h)$

associated with a "world" with its properties. Note that properties and definitions given in this section lead to the implementation presented in Sect. 5.

4.1 Inheritance study

Manipulated combinatorial objects, called elements, are arranged in subsets, called universe, world, family, and sisterhood.

The CalICo universe represents the set of all existing objects and structures in CalICo. Other universes exist, with which CalICo may exchange informations: Maple (Char et al. 1992), Mathematica (Wolfram 1988), $\Lambda\Upsilon\Omega$ (Flajolet et al. 1988), Gaïa (Zimmermann 1994, Flajolet et al. 1994), and Darwin (Bergeron and Cartier 1988) for example. A world is a sub-universe gathering together combinatorial objects that belong to the same class. CalICo's universe is composed of different worlds such as words world, polyominoes world, or trees world. By analogy with statistics, we define quantitative and qualitative predicates. Let P be a predicate on an entity E. If there exists an application, denoted by α from E to Z and an integer n such that

$$\forall x \in E,\ P(x) \implies \alpha(x) \leq n,$$

we say that P is a quantitative predicate. If a predicate is not quantitative, we say that it is a qualitative predicate. Let E be a set of combinatorial objects of the same world and defined by the predicate P. If P is a quantitative predicate and if E is finite then E is a sisterhood, otherwise E is a family. For example, the set of trees verifying the qualitative predicate *to be a binary tree* is a family. The set of binary trees with size equal to 3 is a finite set. It is a sisterhood defined by the quantitative predicate *to have 3 vertices*. Any combinatorial object handled in CalICo is an element.

Remark. In what follows, we will use the term *entity* when we want to refer either to a family or to a sisterhood.

We associate a dependency graph with each world in CalICo. Let E and E' be two entities of the world M, P and P' their respective predicates. We say that E is a parent entity for E' if P' implies P and for any entity E'' in M, with predicate P'', P'' implies P implies P'' implies P'. If E is a parent entity for E', we say that there is a parent–child relation between E and E'. Hence, we define the dependency graph of a world M as a directed labelled graph $G_M = \langle S, A, C\rangle$, where S is the set of entities in M, A is the parent–child relation, and C is the set of predicates defined in M. The transitive closure of the parent–child relation is a relation of order. We call this relation the inheritance relation. At each static step, for one world, we make use of a set of entities linked by the inheritance relation.

Properties of the dependency graph. Let $G_M = \langle S, A, C\rangle$ be the dependency graph of a world M, E and E' be two entities of M, with respective predicates P and P'.

- G_M is a directed acyclic graph – DAG property;
- if $P = P'$ then $E = E'$ – clone property;
- for all path $(s_0, s_1, \ldots, s_l)$ with $l > 1$ in G_M, it does not exist an edge (s_i, s_{i+1}) in S, such that $s_0 = s_i$ and $s_l = s_{i+1}$ – redundance property.

Defining an entity implies new inheritance relationships in the dependency graph. The following information is required: the world where the entity will be classified, the entity's name, a parent entity, an additional predicate, and a set of entities that become the parents of the created entity. The first three parameters are mandatory. One of the latter two is sometimes sufficient. But it is not possible to ignore both at the same time (cf. clone property).

By analogy with the synthesized and inherited attributes that appear in compilers (Aho et al. 1986, pp. 50 and 315), we define the synthesized characteristic and the inherited characteristic. *The membership of an entity for an element* is a synthesized characteristic. This property induces that to obtain the elements of an entity E in the world M one has to collect the elements of E and the elements of all the entities, called E', such that E' is reachable from E in G_M. *To satisfy a predicate* is an inherited characteristic. Parents hand down their predicates to their children.

4.2 Color and combinatorics

Elements brought in CallCo's tutor have a visual representation which is an important feature in their study. The color is used in order to visualize the combinatorial properties. Significance to the colors can be understood by the notions of *attribute, coloring*, and *interpretation* that we introduce below.

Combinatorists enumerate elements according to the value of one or more functions computed on these elements. Those functions are called statistics on elements. In CallCo, a statistic is a function from a family to Z.

Attributes are linked to the drawing of an element. A drawing is geometrically split into indivisible structures. For example the tree drawing needs a circle for each vertex and a segment for each edge. An *attribute* is a type of indivisible structure. Occurrences of attribute are the element's parts that can be colored. For each world, there is a finite number of attributes. For example, a tree has three attributes: vertex, edge, and background. Note that for any object one can put a color on the object's background.

Let us call weight the value of a statistic computed on an element. CallCo associates a color with weights computed on elements. Intuitively, defining a coloring for an element consists in the following actions:

- consider the element on a geometric point of view selecting attributes,
- color the selected attributes according to the weight.

To define formally a coloring, first af all we call $\mathcal{S}$ the set of finite sequences $\mathcal{S} = \bigcup_{k \in N} N^k$. In a world M, we call $\mathcal{D}_M$ the set of pairs (a, s), where a is an attribute and s is in $\mathcal{S}$. Let e be in M, and let $\gamma = \{(a_i, s_i)\}_{i=1,\dots,n}$ be a set of elements in $\mathcal{D}_M$. We say that γ is *a coloring for e* iff:

1. $a_i \neq a_j$ when $i \neq j$,
2. and for all $i = 1, \dots, n$, $|s_i| = n$, where n is either equal to 1 or equal to the occurrences number of a_i in e.

To color an attribute of an element is not sufficient. Attributes are the direct illustration of the element's geometry. We need that the coloring be close to the mathematical properties of the element.

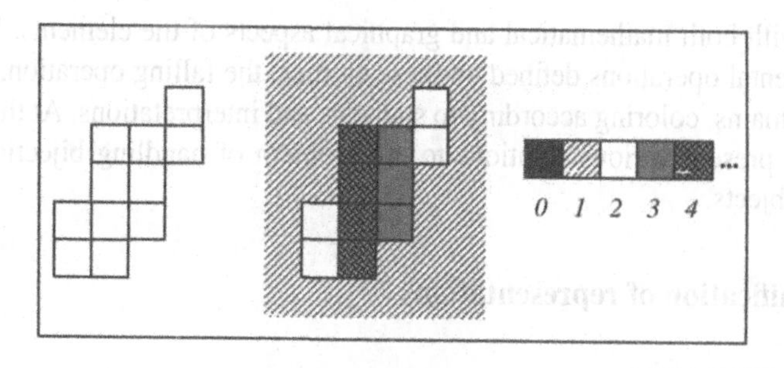

Fig. 9. Example of coloring

For this purpose, we introduce interpretations, which are the links between the geometry and the combinatorics of objects. Formally, an *interpretation* is a function from $(F \times N)$, where F is a family, into the set of possible coloring of an element e in F. Colors are represented symbolically by an integer. Applying an interpretation consists in coloring one or more attributes of e with the given "color" (see the following example and Fig. 9).

In a world M, let $\mathcal{E}_M$ be $\bigcup_{k \in N}(s, \varphi)^k$, where s is either an attribute or an interpretation and φ is a statistic. Note that it is a disjoint union. Let e be an element of M, we define the application, called *color*, from $(M \times \mathcal{E}_M)$ into the set of possible coloring for e.

Example. In Fig. 9, we see two occurrences of a polyominoe. The first one is without coloring. The second one has a coloring which uses two attributes: the cells and the background. The background is hatched if the polyominoe is a parallelogram polyominoe and is black otherwise. (A polyominoe is a *parallelogram polyominoe* if it can be defined by two paths having elementary steps only north or east and having same first point and same last point.) The property "to be a parallelogram" for a polyominoe can be viewed as a statistic. The result, a boolean, is the weight for the polyominoe regarding this statistic. We associate two colors (hatched and black) with this weight. Until now, there is no interpretation. If we want to put the color corresponding to the height of a column on the column, we have to use an interpretation. The statistic is the computation of the height of the column, the result of it is a weight. Setting this weight symbolically represented by a color, on each cell of this column corresponds to the application of an interpretation on the polyominoe.

Remark. Statistics and interpretations are inherited characteristics. Parents hand down their statistics and their interpretations to their children in the dependency graph.

5 Formal coding and symbolic computation

This section concerns the implementation of definitions and properties presented in Sect. 4. We are dealing here with the entities handled in the CalICo tutor: worlds, families, brotherhoods, and elements. First we define a *unique representation* of these entities in order to handle them efficiently. This representation uses formal power series and deals with both mathematical and graphical aspects of the elements. We describe the fundamental operations defined on these entities: the falling operation, recognition of home domains, coloring according to statistics and interpretations. At the end of this section, we present various solutions to the problem of handling bijections between classes of objects.

5.1 Unification of representations

Formal power series

In order to treat correctly the knowledge of different objects, we need a common description of the objects, independent of their world. Formal power series allow us to give unified definitions.

Let us recall that a formal power series S is an application $S: X^* \longrightarrow K$, where X^* is the free monoid over the alphabet X and K a semi-ring. Counting objects using formal power series as generating functions is quite natural.

Definition 1. The generating function associated with a set of objects E according to a parameter size is

$$SG(E) = \sum_{n \geq 0} a_n\, x^n$$

where a_n is the number of elements of E whose size is n.

This definition is classically generalized to an arbitrary number of parameters by associating a variable with each statistic defined on the set E.

Definition 2. Let $\Phi = \{\varphi_1, \varphi_2, \ldots, \varphi_k\}$ be a set of statistics defined on a set of elements E. The generating function of the set E according to the set of statistics Φ is the commutative formal power series associated with E and Φ is defined by

$$SF(E) = \sum_{i_1, i_2, \ldots, i_k \geq 0} a_{i_1, i_2, \ldots, i_k}\, x_1^{i_1}\, x_2^{i_2} \ldots x_k^{i_k},$$

where $a_{i_1, i_2, \ldots, i_k}$ is the number of elements e of E such that $i_j = \varphi_j(c(e))$ for j from 1 to k.

Uniform descriptors

We complete Definition 2 by introducing another alphabet, and more precisely a coding of the objects by means of a formal language. We suppose that there exists a coding c sending each element onto a word. We are now dealing with two alphabets.

- The infinite commutative alphabet $X = x_1, x_2, x_3, \ldots$ is common to all the worlds of the universe and corresponds to the statitics defined in the world.
- Each world M has a proper noncommutative alphabet X_M, eventually infinite and specific to each world. It allows to code the elements of the world M. Let us denote $c(e)$ the word of $X_M{}^*$ coding of an element e.

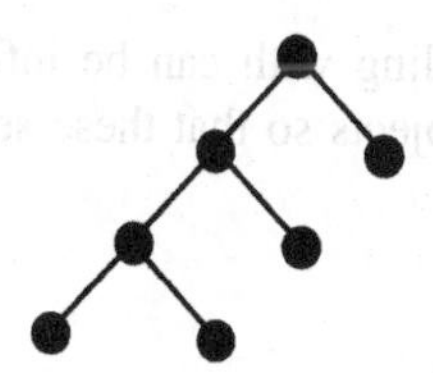

Fig. 10. A binary tree a_0

Definition 3. Each element e of a world M is associated with the following monomial, called descriptor, belonging to the set of the formal power series over X_M with coefficients in $B[x_1, x_2, \ldots, x_k]$,

$$d(e) = \left(\prod_{i=1}^{k} x_i^{\varphi_i(c(e))}\right) c(e).$$

The variables x_i $(1 \leq i \leq k)$ are associated with the statistics φ_i defined on the world M.

Note that the descriptor of an element describes completely the element e. The noncommutative part corresponds to its coding, while the commutative part corresponds to the values of the different statistics defined on M, which is the combinatorial part of e.

Example. Let us give the example of a descriptor of a tree. Let M_a be the world of the trees where $X = \{x_1, x_2, \ldots\}$, $X_{M_a} = \{x, \bar{x}\}$ and a an element of this world. Let us consider the statistics $\varphi_1, \varphi_2, \varphi_3$, associated respectively with the variables x_1, x_2, x_3 such that

- $\varphi_1(a)$ is the depth of the tree a,
- $\varphi_2(a)$ is the number of leaves of a,
- $\varphi_3(a)$ is the number of internal nodes of a.

Let c be the classical coding of binary trees by Dyck words (or well parenthesed systems). For instance, the descriptor of the binary tree a_0 of M_a displayed in Fig. 10 is

$$d(a_0) = x_1^3 x_2^4 x_3^3\, xxx\bar{x}\bar{x}x\bar{x}\bar{x}xx\bar{x}\bar{x}.$$

Definition 3 can be easily extended to descriptors of sisterhoods, families, and worlds by taking formal sums.

Definition 4. The descriptor $SF(S)$ of a set of elements S is defined by the formal sum:

$$SF(S) = \sum_{e \in S} d(e).$$

Note that such descriptors are formal power series over X_M with coefficients in $B[x_1, x_2, \ldots, x_k]$. Thus the descriptors of a brotherhood C, of a family F, of a world M are respectively defined by $SF(C) = \sum_{e\in C} d(e)$, $SF(F) = \sum_{e\in F} d(e)$, $SF(M) = \sum_{e\in M} d(e)$.

Remark. In theory, the formal power series we are dealing with can be infinite. In practice, we are only manipulating a finite number of objects so that these series are always polynomials.

5.2 Operations on the elements

We present here the most interesting operations implemented at the tutor level. We insist on the fact that one can use the tutor in order to study elements that are yet unknown at the tutor level.

Remark. A first version of the tutor had been developed using Chez-Scheme (Rouillon 1991) which is an implementation of Lisp. Then the tutor has been developed in Maple (Char et al. 1992) for many reasons. Firstly, Maple is a complete programming language and a powerful computer algebra system, so that it is well adapted to our formal manipulations. Secondly, the scientific community interested in CalICo generally knows Maple well, and a lot of such scientists have developed their own Maple packages for their research. Thus, using Maple for implementing the tutor gives us the guarantee of a true homogeneity and a good extensibility.

The falling operation

The idea of this operation is to let an element go through entities using the parents–children relation. The predicates associated with each entity have the role of a filter. In order to well understand this falling principle and the other operations, we need to define the notion of home domains of an element.

Definition 5. Let e be an element of a world M. An entity E (family or sisterhood) is a home domain for the element e when:

- e belongs to E and
- e does not belong to any descendant of E in the dependency graph of M.

The generating family of M is the unique home domain of e when e does not belong to any other entity of the world.

Every falling element is found in its home domains and their ascendants.

Example. Consider tree 12 in the world of the trees. It is falling from the generic family of the world, successively passing through the family of the binary trees, then through the family of the complete binary trees, and at last in the brotherhood of the trees having size 7.

Recognition

When manipulating elements, one wants to know its characteristics and properties. Some are quite obvious, others are harder to find. In particular, the knowledge of the different home domains is particularly interesting.

Recognizing an *object* consists in applying the falling operation, that is finding all the home domains containing the element.

Recognizing a *set of elements* consists in finding the set of the home domain which contains all the elements of the set.

Statistics and interpretations

When searching statistics or interpretation for an object, two cases can occur. On the one hand, one wants to know the statistics and interpretations for a known element, that is to say, recorded elements of the tutor. On the other hand, one wants to know the statistics and interpretations for new elements, that is to say, nonrecorded elements of the tutor. In the first case, it is immediate: the information is stored in the tutor. In the other case, one has first to apply the falling principle and then compute the statistics and interpretations according to the home domains. Recall that some statistics can be defined for some families and not for others.

Coloring

The aim is to compute a coloring according to statistics or/and interpretations. First, one has to find the statistics and interpretations as explained before. Then, according to the statistic/interpretation, it is necessary to construct the coding of the object in order to send it to the corresponding visualization workshop.

Conception

The conception of an element can be done either by a visualization workshop or using a generating workshop. It consists essentialy in constructing its coding in the world and to label it by its name. Then the new elements can be recorded in the tutor.

Recording

Recording a set of elements consists first in recognizing the home domains of each of these elements. The recording of several elements is done sequentially. The elements are falling and are recorded successively.

The universe

The main operations concerning the elements have been defined. We can also define new worlds, new families, and new sisterhoods (Rouillon 1994). One of the aims of the tutor is also to check and complete the validity of such new definitions.

5.3 Diffusion of knowledge

The element recorded in a world has been presented up till now from a descriptive point of view. Yet, the recording of an element has implications for the rest of the CalICo universe. Indeed, beyond the data saving, it classifies them with respect to the properties which they verify. As a consequence, the recording of elements enriches other knowledge areas in the tutor.

Here we illustrate this phenomenon by two examples (see Rouillon 1994, for more details).

1. Saving an element e in a family F enriches the family F. So, if the statistic φ applies to elements in F then $\varphi(c(e))$ will be computed and automatically used in all the following requests. We have thus enriched the knowledge on e, on φ, and on F.

2. Let F_1 and F_2 be two distinct families of a world M, such that F_1 (respectively F_2) does not belong to the ascendants of F_2 (respectively of F_1). These hypotheses imply that in the dependency graph G_M of M does not exist a path connecting F_1 and F_2. Then consider φ_1 (respectively φ_2) the unique statistic defined on F_1 (respectively F_2) which is associated with the formal variable x_1 (respectively x_2). Let us denote by S_1 (respectively S_2) their associated descriptors. Suppose that one saves the element e in the world M and that its home domains are F_1 and F_2. Then by definition

$$d(e) = x_1^{\varphi_1(c(e))} x_2^{\varphi_2(c(e))} c(e).$$

We can deduce that the formal power series associated with F_1 and F_2 are modified as follows:

$$d(F_1) = S_1 + x_1^{\varphi_1(c(e))} x_2^{\varphi_2(c(e))} c(e),$$

$$d(F_2) = S_2 + x_1^{\varphi_1(c(e))} x_2^{\varphi_2(c(e))} c(e).$$

There is a distribution of the knowledge inside a world. Indeed, the family F_1 has enriched with the value of the statistic φ_2 perfectly valid on one of its elements while this statistic has never been declared for this family.

Remark. In the next version of CalICo's tutor, we envisage in such a case to introduce the new family satisfying the predicate P, where P is the conjunction of the predicates F_1 and F_2.

5.4 Bijections

We have said that the color in CalICo is used for suggesting the bijections between subsets of worlds. To discover a bijection from a set E_d to a set E_a corresponds to finding a function of translation from the coding in the set E_d to the coding in the set E_a.

Property 6. Bijective relations between worlds can be represented by deterministic finite state automata in the CalICo universe. The transitions of the automaton are the bijections and the states are sisterhoods. We call it world automaton.

Example. Figure 11 shows a world automaton. The worlds are pictured by a very simple element. Note that one cannot always pass from one element to another by transitivity because the bijections are defined for particular classes of elements in each world. Nevertheless, it is clear that the complete automaton for all the bijections induced a transitive relation.

In view of suggesting bijections automatically, three approaches have been proposed in Rouillon (1994):

- to compare sisterhood descriptions among themselves,
- to visualize the modification after each modification of an element,
- to use objects grammar (Dutour and Fédou 1994).

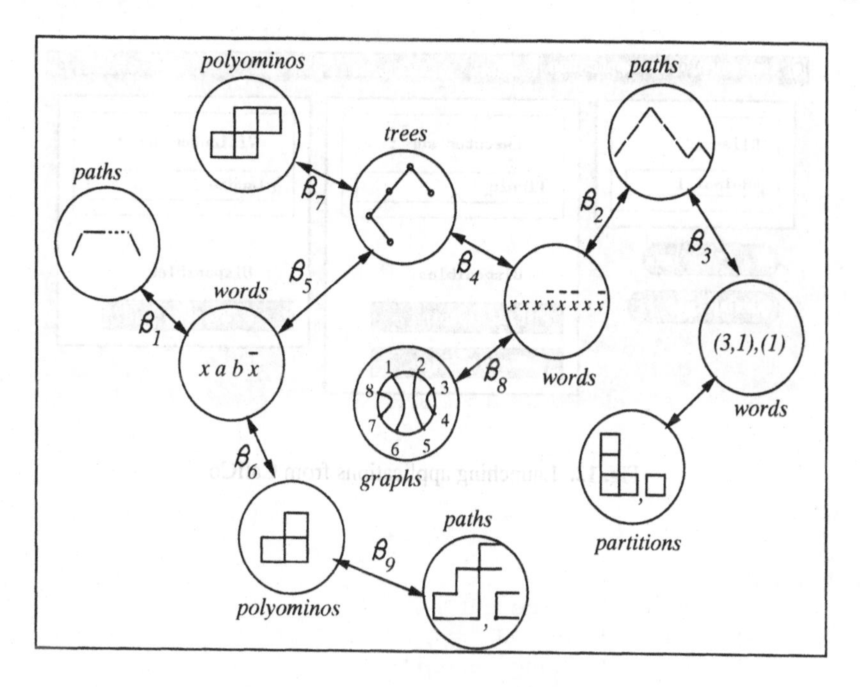

Fig. 11. Example of a world automaton

6 Graphical workshops and graphical interfaces

In this section we will describe CalICo's graphical interfaces. First we will present the piloting interface, called xgeci, which is launched when the `calico` command is issued. Then we will describe the visualization workshops which are available by now in the CalICo environment. We use the terminology of *graphical workshops* for those applications which have nontrivial functionalities. *Graphical interfaces* will denote pure graphical applications which only interface another application. For instance xgeci is a graphical interface because it is only a nice graphical interface for the communications manager GeCI in local mode.

6.1 The piloting interface

In order to conduct a working session with several workshops, CalICo provides a main graphical interface, xgeci (see Fig. 2 a, b). It allows to launch different workshops and to establish communications between them. Thus the user can construct a *virtual machine* (Geist et al. 1993). Each application can be run on different machines and visualized on different displays.

The interface xgeci allows to launch visualization workshops, symbolic computation workshops, or any other programs which have been declared in the configuration file `.CalICo/.CalICo.config`. The possible applications are shown on the left side of the window of xgeci. Clicking an application produces a new dialog box (see Fig. 12). You can specify there on the one hand the machine on which you want a run of the application and on the other hand the display where you want to visualize it.

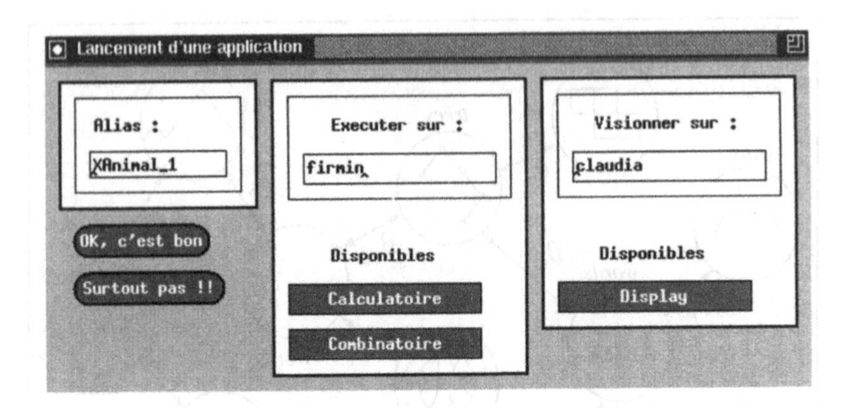

Fig. 12. Launching applications from CallCo

```
APPLICATIONS {
        Nom Interne: "Xpermut"
        Communication: "OUI"
        Fonction: "IGLance_appli"
        Automatique: "NON"
        Machine: "claudia"
        Display: "audrey"
        Arguments: "-pipes","__ID_PIPES__" }

GROUPE MACHINES HOTES {
        "Combinatoire": "claudia","nastasia"
        "Calculatoire": "hector","labri"}

GROUPE MACHINES DISPLAY {
        "Visualisation": "mylene","sharon","audrey"}

CHEMINS {
        Machines: "labri" {
            "Xpermut": "/home/labri/calico/Ateliers/xpermut"}

        Machines: "claudia","nastasia","hector" {
            "Xpermut": "XPermut" } }
```

Fig. 13. Part of a French CallCo configuration file

The different machines and displays proposed by CallCo are those which have been declared in the configuration file. Notice that adding a few lines in this file allows any application to be activated by CallCo. A part of a CallCo configuration file is shown in Fig. 13.

The main graphical interface xgeci enables also to set the communication channels between all these workshops. The communication tools are shown on the right side of the xgeci window. Once you have run applications, you can specify how these applications will communicate by clicking on the single/double/destroy icon and then clicking on the source/target application (see Fig. 2 b). Note that applications or communication channels can be added or removed at every moment of a CalICo session.

6.2 Visualization and manipulation workshops

CalICo offers independent visualization workshops for different kinds of combinatorial objects. Combinatorial objects and their visualization workshops are summarized in Table 1. Visualization workshops have been developed thanks to students' projects. Since 1990, each year one or more groups of three or four students develop new visualization workshops or update old ones. All these workshops have the same philosophy and are somewhat alike even if they are dealing with different kinds of objects. For instance they propose systematically a color map toolbox and a communication box. We will describe them below.

Generic interface

Since 1993, we have unified the implementations of new workshops by an object-oriented approach. Thus, using C^{++} and the Motif library, we developed a model for the visualization workshops, called xmodel. We are dealing now with two kinds of workshops, early ones and xmodel-based ones. By now, there are visualization workshops for polyominoes and animals (*xanimal*), braids (*xtresses*), permutations (*xpermut*), 2-D heap of pieces (*xempil*), graphs (*xgraph*), paths (*xchemin*), and trees (*xarbre*). They all are xmodel-based interfaces except *xtresses*.

Actions such as saving/recovering of objects, copying/pasting/cutting an object, etc. are independent from the nature of objects. These actions are accessible from the main window which is common to all visualization workshops (see Fig. 5). The common window also enables to deal with the colormaps of the visualization workshops. Moreover, the connection tools (send and receive) are also handled by xmodel. Concerning the menus, xmodel uses a language file which contains the name used in the different menus of the visualization workshops. For the moment these menus are all in French, but they can be translated using this language file. The choice of the French language was done because a large part of the community who is dealing with enumerative combinatorics understands French.

Main interface menus

The different menus of the main window of the visualization interfaces (Fig. 5) allows basic operations on the combinatorial objects. In the menu *Fichier* one can find the functions "nouveau" for creating a new drawing area, "charger", "enregistrer" for loading/saving objects, "imprimer" for printing. In the menu *Edition* are the copy/past functions. The painting icon allows to open the colormap and its tools.

Colormaps and colormap tools

One important feature of CalICo is the use of colors to highlight properties of combinatorial objects. Every visualization workshop comes with a colormap and colormap tools.

Table 1. CaICO visualization workshops

Combinatorial objects	Graphical representation	Visualization workshops
Polyominoes		xpolyo (V 3.0) 1990-1992 C, Glib, Graffiti
Permutations		xpermut (V 1.0) 1992 C,AthenaWidget xpermut (V 2.0) 1994 C++, Motif, xmodel
Braids		xtressesimple (β version) 1991 C, Glib, Grafitti xtresses (V 1.0) 1993 C++, AthenaWidget
Heaps of pieces		xempil (V 2.0) 1995 C++, Motif, xmodel
Animals		xanimal (V 1.0) 1995 C++, Motif, xmodel
Trees		xarbres (V 1.0) 1995 C++, Motif, xmodel Knvas widget
Paths		xchemins (V 1.0) 1995 C++, Motif, xmodel

The colormap is a collection of 256 small colored squares, each color corresponds to an integer from 0 to 255. Using the colormap tools enables to transform the default colormap, save or load a colormap. For instance, one can swap two colors of the colormap, execute a circular permutation between several colors, generate a range of hues, create a brand new color by RVB, etc.

Toolbox

Each specific visualization workshop has its own toolbox. It allows to draw objects with the mouse and to modify them (e.g., symmetries and their planar transformations). As an example, the animal/polyominoes toolbox is shown in Fig. 4. Geometric constructions or transformations and algebraic operations are performed on the object(s) with buttons of the toolbox. Each object lives in its own window; these are then itemized in the workshop's main window for special purposes (such as selection).

Mailbox

A mailbox is included in each visualization workshop. When invoking the mailbox, selected objects are sent through a channel that was previously established. More precisely, the workshop sends the formal coding of an object in the communication channel defined at the xgeci level. When a visualization workshop receives a packet, a message is printed in the message window. Clicking the mail box makes the object appear as defined in the preference file (the new object can appear in a new set of objects or in the current set of objects).

6.3 Random generation workshops

Random generation methods have several applications particularly in providing samples of data for evaluation and tests of software and algorithms. Coupling random generation workshops with visualization tools increases considerably the benefit of such tools. At the moment, CalICo environment includes two random generation workshops. One concerns directed animals. The other deals with several types of plane paths, each one based on a specific algorithm (see, e.g., Barcucci et al. 1992, Denise 1993, Wilf 1977).

7 Communication manager

The objective of this section is to explain the construction of a distributed real-time system in our context. The communications between the workshops of CalICo are punctual and nonmassive, and react to a request issued by the user. For an optimal utilization, we have written a graphic interface, called piloting interface allowing a non-specialist of distributed systems to create a virtual machine on which she can work.

Cooperative and real-time distributed systems are very complex to realize, problems of order and coherence being especially difficult to manage. In our system, this management of the order and the coherence has to be carried out by the user. In fact, the information is sent from one workshop to another on the order of a user. Bensoussan et al. (1992) underline that an important research subject consists in simplifying the implementation and the transparency of distributed systems. In this sense, we propose a solution for a particular case. The interactive communication manager, GeCI, is the module built to manage the system of workshops distributed on distant machines and connected to the Internet or to an Ethernet local network. GeCI offers the following properties: the aptitude to add easily a new workshop, the modification of workshop connection during a session, and especially a roundup of several tools specialized in a single software environment. One finds this same approach in a lot of recent projects, such as Meta2-Ψlab (Gomez and Goursat 1994, ΨLab Group 1994), CAS/PI (Kajler 1993), ACELA project (see Cohen and Marteens's contribution to this volume), POSSO (J. Rouiller, oral commun.). A technical advantage of such an approach is that the tools can be developed and maintained independently.

7.1 Design

The interactive communication manager, GeCI, is a partially centralized module that manages the different workshops and creates the cohesion among them.

The purpose is to combine applications to increase their potentials by distributing them on several machines.

Remark. The utilization of UNIX scripts is an example of this method. The juxtaposition of small simple programs allows to realize sophisticateed works.

Overall, our system is an open and extensible system, in which it is essential not to suppress any part of the workshops' functionalities. Three properties are important: the architecture of the system is distributed, each workshop is capable to lead to a result in total isolation from the others (they are thus independent from each other), finally workshops have the possibility to exchange informations.

Connections established between workshops are not static. They are managed dynamically under the user's (or users') control. Users decide also about the transit of the information. The strategy of execution of the work depends therefore exclusively on them.

GeCI establishes a bilateral connection between applications (by utilization of pipes and sockets installed on the Internet domain). It employs the usual primitive of the UNIX operating system and the TCP/IP protocol (Rifflet 1991). By the undertaken technical choice and the installation of an adequate protocol, our communications are reliable and guarantee the arrival of messages in one and only one copy. The communication is by default asynchronous (it is the mode used by CalICo). A request emitted by a workshop does not imply the freezing of this workshop waiting for a reply. The workshop can continue to work and it will detect the reply to its request only when the former will appear on its communication port. However, the synchronous communication mode is possible and is based on the "rendez-vous" principle. To launch an application on a distant machine, one employs a daemon (Rifflet 1991, pp. 215 f), called `calicod`. This daemon is private to CalICo. The client, GeCI, formulates a specific request to the server (`calicod`). Then the daemon executes the service requested: the launching of the wanted application on the remote computer designated. The use of the usual tools TCP/IP and the socket mechanisms guarantee the utilization on all sites.

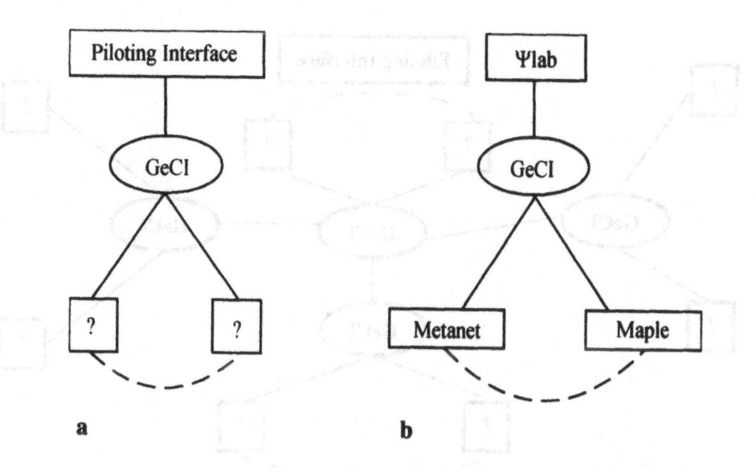

Fig. 14. Diagrams of the two systems integrating GeCI, **a** CallCo, **b** Ψlab

7.2 Design for reusability

An important purpose of the design activity is to allow for the product to evolve and to be used again in others frameworks.

This communication manager has been used in other projects (Gomez et al. oral commun). Figure 14 shows diagrams of the two systems integrating GeCI: CalICo and Ψlab (ΨLab Group 1994).

7.3 System of independent workshops

To anticipate an architecture distributed particularly flexible and based on simple concepts is to anticipate the future. One wishes that a maximum of computers is able to become possible components for the virtual computer of our system. Our system is a partially centralized system with a star-like structure.

Figure 15 gives an idea of the distributed system. Each computer that is a component of the virtual computer has a central module, GeCI, that manages local communications. Each of these modules is connected to a central module, called local GeCI. The local GeCI is in direct collaboration with the user and manages the totality of the system. To launch a workshop on a remote computer, it is useless to know the port number of the computer. One uses the daemon `calicod`. This daemon behaves like the program `rexd` (the SUN RPC server), with the difference that it is less powerful than `rexd` because specific to CalICo but by that less "harmful" for the site security.

For each X application for which the user wants to modify the code in order to make a workshop, one can employ a high-level library of C functions that we have conceived, called CalCom. This library facilitates the integration of an application in the system and provides the communication primitives following our protocol. No management on the sockets has to be made for this application to become a workshop.

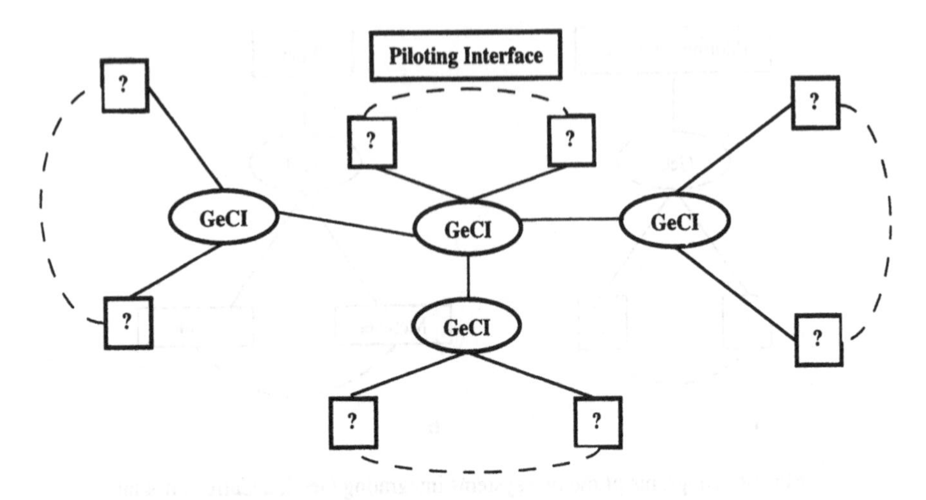

Fig. 15. Set of independent and distributed workshops

7.4 Communications and protocol

Exchanged messages are in the ASCII format. We lose a bit in transmission time during the transit of the message but do not find problems in the framework of the heterogeneous environments (Beaudoin-Lafon and Karsenty 1992). As we said above, we have a specific protocol to our environment that follows a unique principle for all applications integrated to the environment (invisible if one employs the functions of the library CalCom).

Remark. In the framework of CalICo, combinatorial objects that are in transit in the system have a perfectly determined coding. Each application understands this coding and knows how to process it.

Each message is structured in different fields: beginning of message, identifier of sender, identifier of receiver, type of the message, data, end of message. If the workshop does not know which is the address for the message (it is the usual case), it sends its message to GeCI that knows the address of the destination thanks to the connections defined by the user.

The underlying protocol is TCP which provides a local point-to-point order that is sufficient in our case. A limitation of our system resides in the fact that management of the fault tolerance is not complete. However, we manage some chess. If one of the workshops disappears abnormally (e.g., through misfunction or crash of a computer), the system is refreshed automatically. The virtual machine is updated and the graphic piloting interface suppresses the concerned workshop in the central part of its window.

7.5 Communications by copy/paste

Another library of C functions, called libXSel, allows the transit of data between graphic applications by copy/paste with the mouse. The utilization of functionalities of this library has been presented in Sect. 3. However, this particularly convivial mechanism does not allow the control on data and does not allow to employ screens managed by different X servers. The implementation uses the CUT-BUFFERS mechanism of X11 to forward information.

7.6 Integration of an interpreter in the system

The encapsulation is mandatory for some tools if sources are not available. We have been interested especially in this problem with respect to Maple. We propose a generic interface, GeCA, that is usable for all types of interpreter programs (Maple, Scheme, etc.). To be capable to communicate with the interpreter in a transparent manner for the user (that is by executing in background and by diverting standard input/output) is not sufficient. We want to work in interaction with this interpreter, as it is usual to make it but in addition by having the possibility to make it communicate. We want to make a full-fledged workshop with which the user will be able to work directly. This interface has therefore to manage *two inputs*, the keyboard and the communication with the external world, and *two outputs*, the screen and the communication with the external world.

This interface of communications between the interpreter and "the external world" allows the interpreter to work without losing its usual qualities and its functionalities. Furthermore, it offers specific communication functions. For a precise description of its implementation, see Rouillon (1994).

8 Availability and implementation

An experimental version of CalICo is available by mail at calico@labri.u-bordeaux.fr.

CalICo runs on workstation SPARC, under UNIX system, with an environment X-Window (X11 R4 for instance). It is not indispensable, but it is strongly counseled to use a color screen. The installation requires 8 Mo disk space. It is not necessary to have an important central memory or a large zone of swap for an ordinary utilisation. Table 2 gives the memory used for a typical session with five workshops.

GeCI version 1.1 has been carried out in collaboration with the members of the

Table 2. Allocated memory for a typical CalICo session

Memory in Ko	Application
68	`geci`
576	`xgeci`
548	`xpolyo`
452	`xtresse`
616	`xpermut`
76	`geca`
344	`mapleV`
580	`xg-polyo`
3260	Total for the CalICo environment

project META2 of INRIA-Rocquencourt on the following equipments: SPARC SUN4, HP 9000, IBM RS 6000, DEC ALPHA, DEC MIPS, PC with LINUX.

Currently the binaries that compose the CalICo environment are the following.

- `calicod`: this daemon process receives run orders from the client for remote applications.
- `geci`: the interactive communications manager holds two modes. (1) Local mode: `geci` is running on the machine of the user. Joined with the piloting workshop `xgeci`, it executes orders of the user regarding launching remote workshops and interconnecting them. (2) Remote mode: the set of all the machines defined by the user constitute a virtual machine. On each of these machines turns an occurrence of the process `geci`.
- `xgeci`: the graphic piloting interface of `geci` allows a user to launch and suppress interactively the workshops. It allows to create and to destroy connections between them. The interface `xgeci` transmits to `geci` all the operations to do.

Other interfaces exist (new workshops come frequently in the CalICo environment): `xtuteur` the tutor graphic interface, `simulcra` a communication debugger for CalICo, `geca` communication interface for interpretors, `xpolyo` for the polyominos, `xtresse` for braids, `xpermut` for permutations, `xempil2D` for 2-D heap of pieces.

A collection of workshops for the random generation of combinatorial objects are available.

9 Conclusion

We have shown CalICo as a software system intended for combinatorial researchers. Thanks to GeCI, in a distributed environment the user handles combinatorial objects, both in a graphical and a mathematical way. Having a consistent environment that integrates graphical programs, symbolic-computation programs, and a knowledge-based system, CalICo is a powerful tool which helps users to gain better insight into combinatorial problems. Graphical programs offer the possibility to draw objects and symbolic-computation programs allow to work on their associated formal coding. A special tool, the tutor, written in Maple, classifies mathematical objects according to their combinatorial properties. In this approach, properties are automatically collected for each manipulated object. The color is a main component in CalICo: it is the visual support that transcribes the computed statistics on objects. Of course the main remark can be that the number of workshops is small and especially oriented towards the objects that we are studying frequently in our laboratories in Bordeaux. Thus a future work would be to implement new workshops by the xmodel generator. Another goal will be to experiment with CalICo in order to know if the tool is powerful enough for finding new bijections. Furthermore, we also want to explore CalICo's use for electronic publishing.

Acknowledgments

CalICo project is partially supported by EC grant CHRX-CT93-0400, PRC-Maths/Info and GDR Programmation.

References

Aho, A., Sethi, R., Ullman, J. (1986): Compilers. Addison-Wesley, Reading, MA.

Arques, D., Eyrolles, G., Janey, N., Viennot, X. (1989): Combinatorial analysis of ramified patterns and computer imagery of trees. ACM SIGGRAPH. Comput. Graph. 23/3: 31–40.

Barcucci, E., Pinzani, R., Sprugnoli, R. (1992): Génération aléatoire de chemins sous diagonaux. In: Leroux, P., Reutenhauer, C. (eds.): Actes du 4ème Colloque Séries Formelles et Combinatoire Algébrique. Publications LaCIM, 11, Université de Québec à Montreal, pp. 17–32.

Beaudoin-Lafon, M., Karsenty, A. (1992): Transparency and awareness in a real-time groupware system. In: ACM Symposium on User Interface Software and Technology UIST '92, Association for Computing Machinery, New York, pp. 171–180.

Bensoussan, A., et al. (1992): Les grands sujets de recherches en informatique et les programmes de l'INRIA. 25^{ieme} anniversaire de l'INRIA. Institut National de Rechercheen Informatique et en Automatique, Le Chesnay.

Bergeron, F., Cartier, G. (1988): Darwin: computer algebra and enumerative combinatorics. In: Cori, R., Wirsing, M. (eds.): STACS '88. Springer, Berlin Heidelberg New York Tokyo, pp. 393–394 (Lecture notes in computer science, vol. 294).

CallCo Group (1993): Outil de développement pour les communications, manuel d'utilisation. Rapp. Techn., LaBRI, Univerite de Bordeaux I, Bordeaux, France.

Char, B., Geddes, K., Gonnet, G., Leong, B., Monogan, M., Watt, S. (1992): First leaves: a tutorial introduction to Maple V. Springer, Berlin Heidelberg New York Tokyo.

Delest, M. (1995): Algebraic languages: a bridge between combinatorics and computer science. In: Billera, L. J., Greene, C., Simion, R., Stanley, R. P. (eds.): Formal power series and algebraic combinatorics. American Mathematical Society, Providence, RI, pp. 71–88.

Denise, A. (1993): Génération aléatoire et uniforme de mots. In: Barlotti, A., Delest, M., Pinzani, R. (eds.): Actes du 5ème Colloque Séries Formelles et Combinatoire Algébrique, 1993. Université de Florence, pp. 153–164.

Denise, A., Rouillon, N. (1992): Génération de structures arborescentes. In: Lucas, M. (ed.): Actes des 5 journées GROPLAN, pp. 83–90.

Dutour, I., Fédou, J. (1994): Grammaire d'objets. Tech. Rep. 963-94, LaBRI, Université de Bordeaux I, Bordeaux, France.

Flajolet, P., Salvy, B., Zimmermann, P. (1989): $\Lambda\Upsilon\Omega$: an assistant algorithms analyser. In: Mora, T. (ed.): Applied algebra, algebraic algorithms, and error-correcting codes. Springer, Berlin Heidelberg New York Tokyo, pp. 201–212 (Lecture notes in computer science, vol. 357)

Flajolet, P., Zimmermann, P., Cutsem, B. V. (1994): A calculus for the random generation of labelled combinatorial structures. Theor. Comput. Sci. 132: 1–35.

Gaudin, V. (1995): XModele 2.0: manuel de maintenance. LaBRI, Université Bordeaux I, no. 9994–95.

Geist, A., Beguelin, A., Dongarra, J., Jiang, W., Manchek, R., Sunderam, V. (1993): PVM 3 users's guide and reference manual. Techn. Rep., Oak Ridge National Laboratory, Oak Ridge, TN.

Gomez, C., Goursat, M. (1994): Metanet: a system for network analysis. DIMACS Ser. Discr. Math. Theor. Comput. Sci. 15: 255–268.

Kajler, N. (1993): Environnement graphique distribué pour le calcul formel. Ph.D. thesis, Université de Nice-Sophia Antipolis, Sophia Antipolis, France.

Kerber, A. (1991): Algebraic combinatorics via finite group actions. B. I. Wissenschafts verlag, Mannheim.

ΨLab Group (1994): Ψlab user's guide. Institut National de Recherche en Informatique et en Automatique, Le Chesnay.

Rifflet, J. (1991): La communication sous Unix. Ediscience, Paris.

Robinson, G. D. B. (1948): On the representation of the symmetric group. Am. J. Math. 60: 745–760

Rouillon, N. (1991): CalICo: une première réalisation du noyau. Mémoire de DEA Informatique, Université Bordeaux I, Bordeaux, France.

Rouillon, N. (1994): Calcul et image en combinatoire. Ph.D. thesis, Université de Bordeaux I, Bordeaux, France.

Schensted, C. (1961): Longest increasing and decreasing subsequences. Can. J. Math. 13: 179–191.

Skiena, S. (1990): Implementing discrete mathematics: combinatorics and graph theory with Mathematica. Addison-Wesley, Reading, MA.

Symbolics (1984): Macsyma reference manual, 3rd edn. Symbolics Inc., Cambridge, MA.

Viennot, G. (1986): Heaps of pieces i: basic definitions and combinatorial lemmas. In: Labelle, G., Leroux, P. (eds.): Combinatoire enumérative. Springer, Berlin Heidelberg New York Tokyo, pp. 210–245 (Lecture notes in mathematics, vol. 1234).

Viennot, X. (1977): Une forme géométrique de la correspondance de Robinson-Schensted. In: Foata, D. (ed.): Combinatoire et représentation du groupe symétrique. Springer, Berlin Heidelberg New York, pp. 29–58 (Lecture notes in mathematics, vol. 579).

Viennot, X. (1988): La combinatoire bijective par l'exemple. Rapp. Interne, LaBRI, Université de Bordeaux, Bordeaux, France.

Viennot, X. (1992): A survey of polyominoes enumeration. In: Leroux, P., Reutenauer, C. (eds.): 4ème Colloque Séries Formelles et Combinatoire Algébrique. Publications LaCIM, 11, Université du Québec à Montréal, pp. 399–420.

Wilf, H. (1977): A unified setting for sequencing, ranking, and selection algorithms for combinatorial objects. Adv. Math. 24: 281–291.

Wolfram, S. (1988): Mathematica: a system for doing mathematics by computer. Addison-Wesley, Reading, MA.

Zimmermann, P. (1994): Gaïa: a package for the random generation of combinatorial structures. Maple Techn. Newslett. 1/1: 38–46.

Erratum

The title page of M. Beeson's contribution, positioned as p. 163, should be p. 89.

The title page of O. Arsac et al.'s contribution, positioned as p. 89, should be p. 163.

Subject index

Texts and Monographs in Symbolic Computation

Bob F. Caviness, Jeremy R. Johnson (eds.)

Quantifier Elimination and Cylindrical Algebraic Decomposition

1998. 20 figures. XIX, 431 pages.
Soft cover DM 118,–, öS 826,–, US $ 79.95. ISBN 3-211-82794-3

George Collins' discovery of Cylindrical Algebraic Decomposition (CAD) as a method for Quantifier Elimination (QE) for the elementary theory of real closed fields brought a major breakthrough in automating mathematics with recent important applications in high-tech areas (e.g. robot motion), also stimulating fundamental research in computer algebra over the past three decades.

This volume is a state-of-the art collection of important papers on CAD and QE and on the related area of algorithmic aspects of real geometry.

In addition to original contributions by S. Basu et al., L. González-Vega et al., G. Hagel, H. Hong and J.R. Sendra, J.R. Johnson, S. McCallum, D. Richardson, and V. Weispfenning and a survey by G.E. Collins outlining the twenty-year progress in CAD-based QE it brings together seminal publications from the area:

A. Tarski: A Decision Method for Elementary Algebra and Geometry
G.E. Collins: Quantifier Elimination for Real Closed Fields by Cylindrical Algebraic Decomposition
M.J. Fischer and M.O. Rabin: Super-Exponential Complexity of Presburger Arithmetic
D.S. Arnon et al.: Cylindrical Algebraic Decomposition I: The Basic Algorithm; II: An Adjacency Algorithm for the Plane
H. Hong: An Improvement of the Projection Operator in Cyclindrical Algebraic Decomposition
G.E. Collins and H. Hong: Partial Cylindrical Algebraic Decomposition for Quantifier Elimination
H. Hong: Simple Solution Formula Construction in Cylindrical Algebraic Decomposition Based Quantifier Elimination
J. Renegar: Recent Progress on the Complexity of the Decision Problem for the Reals

SpringerWienNewYork

Sachsenplatz 4-6, P.O.Box 89, A-1201 Wien, Fax +43-1-330 24 26, e-mail: order@springer.at, Internet: http://www.springer.at
New York, NY 10010, 175 Fifth Avenue • D-14197 Berlin, Heidelberger Platz 3 • Tokyo 113, 3-13, Hongo 3-chome, Bunkyo-ku

Texts and Monographs in Symbolic Computation

Alfonso Miola, Marco Temperini (eds.)

Advances in the Design of Symbolic Computation Systems

1997. 39 figures. X, 259 pages.
Soft cover DM 98,–, öS 682,–, US $ 79.95. ISBN 3-211-82844-3

New methodological aspects related to design and implementation of symbolic computation systems are considered in this volume aiming at integrating such aspects into a homogeneous software environment for scientific computation. The proposed methodology is based on a combination of different techniques: algebraic specification through modular approach and completion algorithms, approximated and exact algebraic computing methods, object-oriented programming paradigm, automated theorem proving through methods à la Hilbert and methods of natural deduction. In particular the proposed treatment of mathematical objects, via techniques for method abstraction, structures classification, and exact representation, the programming methodology which supports the design and implementation issues, and reasoning capabilities supported by the whole framework are described.

Franz Winkler

Polynomial Algorithms in Computer Algebra

1996. 13 figures. VIII, 270 pages.
Soft cover DM 89,–, öS 625,–, US $ 69.00. ISBN 3-211-82759-5

The book gives a thorough introduction to the mathematical underpinnings of computer algebra. The subjects treated range from arithmetic of integers and polynomials to fast factorization methods, Gröbner bases, and algorithms in algebraic geometry. The algebraic background for all the algorithms presented in the book is fully described, and most of the algorithms are investigated with respect to their computational complexity. Each chapter closes with a brief survey of the related literature.

Sachsenplatz 4-6, P.O.Box 89, A-1201 Wien, Fax +43-1-330 24 26, e-mail: order@springer.at, Internet: http://www.springer.at
New York, NY 10010, 175 Fifth Avenue • D-14197 Berlin, Heidelberger Platz 3 •Tokyo 113, 3-13, Hongo 3-chome, Bunkyo-ku

Texts and Monographs in Symbolic Computation

Jochen Pfalzgraf, Dongming Wang (eds.)

Automated Practical Reasoning

Algebraic Approaches

With a Foreword by Jim Cunningham
1995. 23 figures. XI. 223 pages.
Soft cover DM 108.–, öS 755.–, US $ 59.00. ISBN 3-211-82600-9

This book presents a collection of articles on the general framework of mechanizing deduction in the logics of practical reasoning. Topics treated are novel approaches in the field of constructive algebraic methods (theory and algorithms) to handle geometric reasoning problems, especially in robotics and automated geometry theorem proving; constructive algebraic geometry of curves and surfaces showing some new interesting aspects; implementational issues concerning the use of computer algebra systems to deal with such algebraic methods. Besides work on nonmonotonic logic and a proposed approach for a unified treatment of critical pair completion procedures, a new semantical modeling approach based on the concept of fibered structures is discussed; an application to cooperating robots is demonstrated.

Wen-tsün Wu

Mechanical Theorem Proving in Geometries

Basic Principles

Translated from the Chinese by Xiaofan Jin and Dongming Wang
1994. 120 figures. XIV. 288 pages.
Soft cover DM 98.–, öS 686.–, US $ 69.00. ISBN 3-211-82506-1

Bernd Sturmfels

Algorithms in Invariant Theory

1993. 5 figures. VII. 197 pages.
Soft cover DM 65.–, öS 455.–, US $ 41.95. ISBN 3-211-82445-6

Sachsenplatz 4-6, P.O.Box 89, A-1201 Wien, Fax +43-1-330 24 26, e-mail: order@springer.at, Internet: http://www.springer.at
New York, NY 10010, 175 Fifth Avenue • D-14197 Berlin, Heidelberger Platz 3 •Tokyo 113, 3-13, Hongo 3-chome, Bunkyo-ku

Springer-Verlag and the Environment

WE AT SPRINGER-VERLAG FIRMLY BELIEVE THAT AN international science publisher has a special obligation to the environment, and our corporate policies consistently reflect this conviction.

WE ALSO EXPECT OUR BUSINESS PARTNERS – PRINTERS, paper mills, packaging manufacturers, etc. – to commit themselves to using environmentally friendly materials and production processes.

THE PAPER IN THIS BOOK IS MADE FROM NO-CHLORINE pulp and is acid free, in conformance with international standards for paper permanency.